Biology of the Indian Stingless Bee:

Tetragonula iridipennis Smith

Biology of the Indian Stingless Bee:

Tetragonula iridipennis Smith

By

Ujjwal Layek and Prakash Karmakar

Cambridge Scholars Publishing

Biology of the Indian Stingless Bee: *Tetragonula iridipennis* Smith

By Ujjwal Layek and Prakash Karmakar

This book first published 2025

Cambridge Scholars Publishing

Lady Stephenson Library, Newcastle upon Tyne, NE6 2PA, UK

British Library Cataloguing in Publication Data
A catalogue record for this book is available from the British Library

ISBN: 978-1-0364-4303-0
ISBN (Ebook): 978-1-0364-4304-7

CONTENTS

PREFACE

With evolution, both plant and animal kingdoms have developed into diverse species. Plants and animals interact in different ways, with the dominating type being mutual interactions among flowering plants and their pollinators. The most important pollinators are insects, including honeybees, solitary bees, and stingless bees. Through their pollination services, they help conserve plant diversity and optimise crop yield. Besides ecosystem services, some insects' nests or hive products (e.g., honeybees and stingless bees) are highly demanding and utilised from different perspectives.

Unfortunately, the population of pollinators, including honeybees and stingless bees, is declining worldwide. It may be due to climate change, habitat degradation, extensive monoculture, and the use of agrochemicals. The decline of pollinators negatively impacts biodiversity, crop yield, and, thereby, the economy of farmers. In recent years, beekeepers and researchers have started to manage bees (e.g., honeybees, orchid bees, mason bees, and stingless bees) to overcome the pollination deficit of crops. However, the management of non-native bees negatively influences the foraging behaviour of local pollinators and their survival. So, beekeepers and researchers are focusing on the management of native pollinators.

The Indian stingless bee *Tetragonula iridipennis* Smith, a unique species, is distributed in India, the Indomalayan realm, Nepal, and Sri Lanka. The species is widely distributed in India, from sea level to higher altitudes, including several states. The stingless bees forage on diverse plant species with significant floral fidelity. They are effective pollinators of many crops. However, little information about stingless bees is available, and the data deficit may render meliponiculture sustainable. Thus, comprehensive data about the Indian stingless bees is in high demand.

This book covers almost all aspects (e.g., distribution, morphology, colony organisation, nesting biology, foraging behaviour, threats and conservation strategies, bee products, pollination efficiency, etc.) of the stingless bee species. This information helps readers be aware of the stingless bee that can help conserve the bee species. It is hoped that the book will be useful to beekeepers, researchers, and students in acquiring knowledge about Indian stingless bees, which will help in the scientific management of stingless bees. The book will guide the beekeeper in extracting bee products from a hive. It will generate enough interest among scholars to inspire and guide future research on stingless bees.

Ujjwal Layek
Prakash Karmakar

CHAPTER 1

TAXONOMIC NOTES AND DISTRIBUTION OF INDIAN STINGLESS BEES

Stingless bee species in India

Rasmussen (2013) reported eight species of stingless bees from the Indian subcontinent. These were *Lepidotrigona arcifera* (Cockerell), *Lisotrigona cacciae* (Nurse), *Lisotrigona mohandasi* (Jobiraj & Narendran), *Tetragonula aff. laeviceps* (Smith), *Tetragonula bengalensis* (Cameron), *Tetragonula gressitti* (Sakagami), *Tetragonula iridipennis* (Smith), *Tetragonula praeterita* (Walker), and *Tetragonula ruficornis* (Smith). From North East, North, and South India, Rahaman et al. (2015) reported two genera and six different species of stingless bees: *Lepidotrigona arcifera*, *Tetragonula bengalensis*, *Tetragonula iridipennis*, *Tetragonula laeviceps*, *Tetragonula praeterita* and *Tetragonula ruficornis*. Viraktamath and Roy (2022) described five species of *Tetragonula*, namely, *Tetragonula vikrami* Viraktamath, (from Karnataka), *T. sumae* Viraktamath, (from Tamil Nadu), *T. ashishi* Viraktamath and Jagruti, (from Maharashtra), *T. shishirae* Viraktamath (from Rajasthan) and *T. shubhami* Viraktamath (from Chhattisgarh).

Systematic position of *Tetragonula iridipennis*

The Indian stingless bee (also called dammar bee), *Tetragonula iridipennis*, is a species belonging to the family Apidae, subfamily Apinae and tribe Meliponini. It was first described by Frederick Smith in 1854, who found the species on the island of Sri Lanka (Rasmussen 2013). Many older literature erroneously placed this species in the genus *Melipona*. Until recently, it was classified as *Trigona* and is still mistakenly referred to as *Trigona iridipennis*. The systematic position of the stingless bee is given below:

Domain: Eukaryota
Kingdom: Animalia
Phylum: Arthropoda
Class: Insecta
Order: Hymenoptera
Family: Apidae
Tribe: Meliponini
Genus: *Tetragonula*
Species: *T. iridipennis*

Distribution of Indian stingless bees

The stingless bees are native to the tropical and subtropical regions of the world. Stingless bees are divided into three monophyletic clades, and their distribution today largely corresponds to three major biogeographic regions- Neotropical, African, and Indo-Malayan-Australasian (Rasmussen and Cameron 2010). Indian stingless bee (*Tetragonula iridipennis*) is found in India, the Indomalayan realm, Nepal, and Sri Lanka. In India, the species is widely distributed, from sea level to a higher altitude, including several states like Jannu & Kashmir, Maharastra, Kerala, Tamil Nadu, and West Bengal (Layek and Karmakar 2018; Rahaman 2018; Balaji et al. 2023).

The distribution pattern is clumping to a random type (Figure 1.1). The minimum distance between the nearest two colonies is 5.6 m. The number of colonies per kilometre square area (i.e., density) varied from 0 to 18 (4.3 ± 3.46). Densities are comparatively higher in mixed vegetation (having many trees, shrubs, and herbs) and human habitats (having building structures and plants) than in agricultural field-dominated areas and deep forest areas (Table 1.1). In the Agricultural zone, they may quickly find their preferred nesting substrates, like building walls and tree trunks. In the deep forest area of Sal (*Shorea robusta* L.), nesting cavities may be unavailable. Also, they may not have diverse floral resources available. As their colonies are perennial, colony density does not vary by season, unlike Indian honeybees.

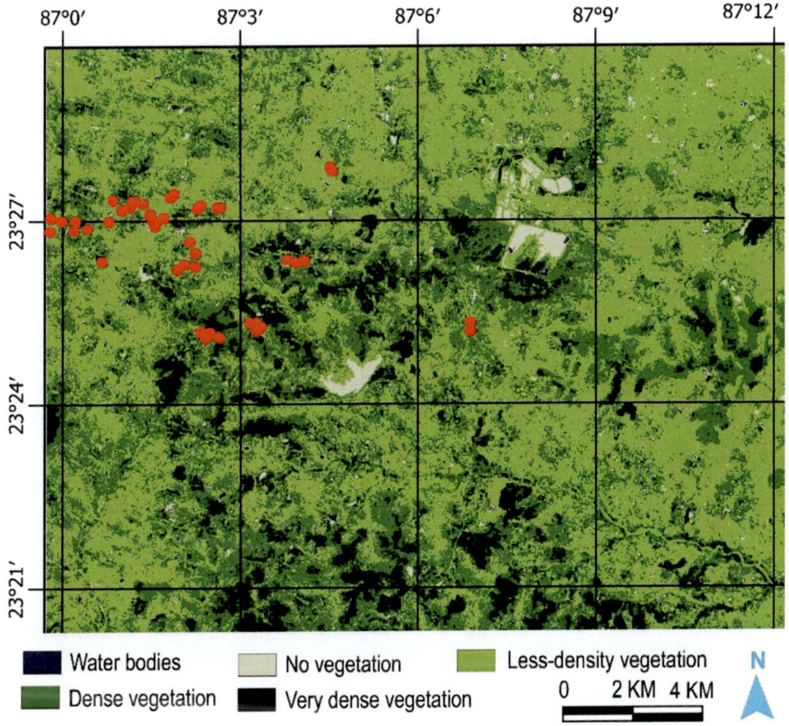

Figure 1.1. Map showing the clumping distribution of stingless bee colonies (red circles).

Table 1.1. Density (number of colonies/ m^2 area) of stingless bees in different landscapes.

Vegetation type	Density
Human settlement without vegetation	$0.2^{d} \pm 0.42$
Human settlement with sparse vegetation	$7.50^{a} \pm 3.31$
Mixed vegetation[1]	$4.90^{b} \pm 2.92$
Agricultural landscape	$0.40^{d} \pm 0.70$
Deep forest (tree dominated)	$1.10^{c} \pm 1.10$
Polluted area (factory dominated)	$0^{e} \pm 0$

[1]Mixed vegetation- moderate dense vegetation with a mixture of a few crop fields, trees, and bushy areas. Values are given in mean ± standard deviation. Different superscript letters (shown after the mean values) indicate significant differences (Dunn's post hoc test at 5%).

Factors influencing the distribution of stingless bee colonies

Stingless bee colonies are not homogeneously distributed throughout the landscape. In some zones (e.g., human settlement areas with sparse vegetation), their density remains higher than in others, like agricultural landscapes and human settlement areas without significant vegetation. Many factors are responsible for their unequal distribution:

- **Availability of nesting substrate:**
The presence of suitable nesting substrates plays a pivotal role in determining the distribution and population density of stingless bee colonies. These bees are essential pollinators within tropical ecosystems and require specific nesting materials, such as hollow tree trunks, branches, or soil cavities, which vary by species. These substrates' characteristics, quality, and quantity in a particular region can significantly influence stingless bee colonies' concentration and spatial arrangement. In areas with appropriate nesting sites, bee populations are generally more dense and widespread. In contrast, regions where nesting substrates are limited due to deforestation, urban development, or agricultural growth experience a fragmented distribution of stingless bee colonies, often resulting in population declines. The competition for scarce nesting resources can intensify this challenge, leading to conflicts among species or compelling colonies to inhabit less-than-ideal sites, which may adversely affect their health and productivity. Furthermore, the presence of natural predators and environmental pressures, including climate change, can alter the suitability of nesting substrates, affecting the overall distribution of stingless bees. Gaining insight into the connection between the availability of nesting substrates and the distribution of stingless bees is vital for conservation initiatives, as it underscores the importance of preserving natural habitats and ensuring the availability of nesting resources to sustain healthy bee populations and the ecosystems they support.

- **Availability of raw nesting materials:**
The presence of raw nesting materials is crucial in determining the distribution and viability of stingless bee colonies. These bees depend on a range of natural substances, including resins, plant fibres, mud, and wax, for the construction and upkeep of their nests. The quantity and accessibility of these materials within an environment can influence

the locations where colonies choose to settle and their ability to thrive and expand. In areas with abundant raw nesting materials, stingless bees are likelier to flourish, resulting in greater densities of colonies. These materials are vital for the initial nest construction and ongoing maintenance and establishing protective barriers against predators and environmental threats. Conversely, regions with limited access to these resources may experience a decline in colony establishment as bees face challenges in building or maintaining their nests, potentially leading to a decrease in local populations. Human activities such as deforestation, agriculture, and urban development can further diminish the availability of raw nesting materials by transforming the landscape and reducing essential resources. This scarcity compels bees to travel greater distances to collect necessary materials, which can increase energy expenditure and negatively impact the overall health and productivity of the colony. Therefore, the distribution of stingless bee colonies is intricately connected to the availability of suitable nesting materials, underscoring the importance of conserving natural habitats and ensuring the availability of these resources for the sustainability of these vital pollinators.

- **Availability of food resources:**

Food resources are vital to stingless bee colonies' distribution and population density. These bees rely on a steady supply of nectar and pollen from flowering plants to nourish their colonies, support their larvae, and produce honey. The variety and quantity of floral resources in a specific area significantly determine the establishment and prosperity of stingless bee colonies. In regions abundant with flowering plants, particularly those that bloom year-round, stingless bees are more inclined to develop dense and stable populations, as they can readily access the essential resources needed for survival. In contrast, in areas where floral resources are limited, seasonal, or severely fragmented due to deforestation, agriculture, or urbanisation, the distribution of stingless bee colonies tends to be more dispersed and less concentrated. The bees may face difficulty locating sufficient food, resulting in longer foraging distances, which can diminish colony efficiency and survival rates. Additionally, competition for scarce floral resources can be fierce, both among different bee species and within the same species, potentially causing some colonies to be displaced. Climate change further complicates this scenario, as alterations in temperature and precipitation can disrupt the flowering patterns of plants, affecting food availability for stingless bees.

Consequently, preserving diverse, flower-rich habitats is crucial for sustaining healthy populations of stingless bees, as these environments provide the essential food resources that directly impact their distribution and ecological viability.

- **Pollution:**

Pollution profoundly affects the distribution and health of stingless bee colonies, presenting a significant threat to their survival and the ecosystems they sustain. Various forms of pollution, including air, water, and soil contamination, often stemming from industrial processes, agricultural practices, and urban development, can affect stingless bees directly and indirectly. The use of pesticides and herbicides in agriculture is particularly detrimental, as these chemicals can taint the bees' food sources—nectar and pollen—resulting in toxic exposure that compromises their foraging behaviour, reproductive success, and overall colony health. Additionally, airborne pollutants can hinder the bees' navigational abilities, as many species depend on chemical signals and pheromones for communication and orientation. This disorientation may lead to difficulty returning to their nests, consequently increasing mortality rates. Water pollution, caused by contaminating water sources with heavy metals or other hazardous substances, poses further risks when bees collect water for their colonies, introducing harmful elements into their living environment. Furthermore, soil pollution can diminish the quality of nesting sites, complicating searching for suitable locations for colony establishment. As pollution alters the availability and quality of environmental resources, it may compel stingless bees to migrate, decrease their population densities, or even result in local extinctions in areas with severe pollution. Therefore, the cumulative effects of pollution disrupt the natural distribution patterns of stingless bee colonies, underscoring the urgent need for environmental protections and sustainable practices to preserve these essential pollinators and the ecosystems they support.

CHAPTER 2

MORPHOMETRY OF INDIAN STINGLESS BEES

There are three casts of Indian stingless bees: workers (sterile females), queens (fertile females), and drones (fertile males). Workers and drones are similar in external morphology; drones are slightly darker than workers. The queen is much larger than workers and drones (Figure 2.1).

Figure 2.1. Different castes of stingless bees. (A) Worker, (B) queen, and (C) drone. Scale bars = 1 mm.

External morphology of a worker bee

A. Head

Head length is 1.38 ± 0.21 mm (1.20–1.55 mm), and width is 1.71 ± 0.12 mm (ranged: 1.60–1.82 mm). Stingless bees have five eyes- two large compound eyes and three smaller simple eyes (also called ocelli). The median ocellus width is 82.63 ± 12.54 µm. The inter-ocellar distance is 176.57 ± 14.82 µm. The ocello-ocular distance is 284.61 ± 23.03 µm. The hexagonal cells in the compound eyes are about 18 µm in length and 16 µm in breadth (Figure 2.2). Inward of the compound eyes, there are many uniparous branched hairs about 25 µm long. The antennal length is 1.82 ± 0.08 (ranges: 1.65–1.81 mm), comprises of many segments. Each segment is about 60–80 µm in length. The antennal surfaces bear many unbranched hairs about 12 μm long (Figure 2.2). The proboscis is about 1.2 mm. The inner tube is about 850 μm.

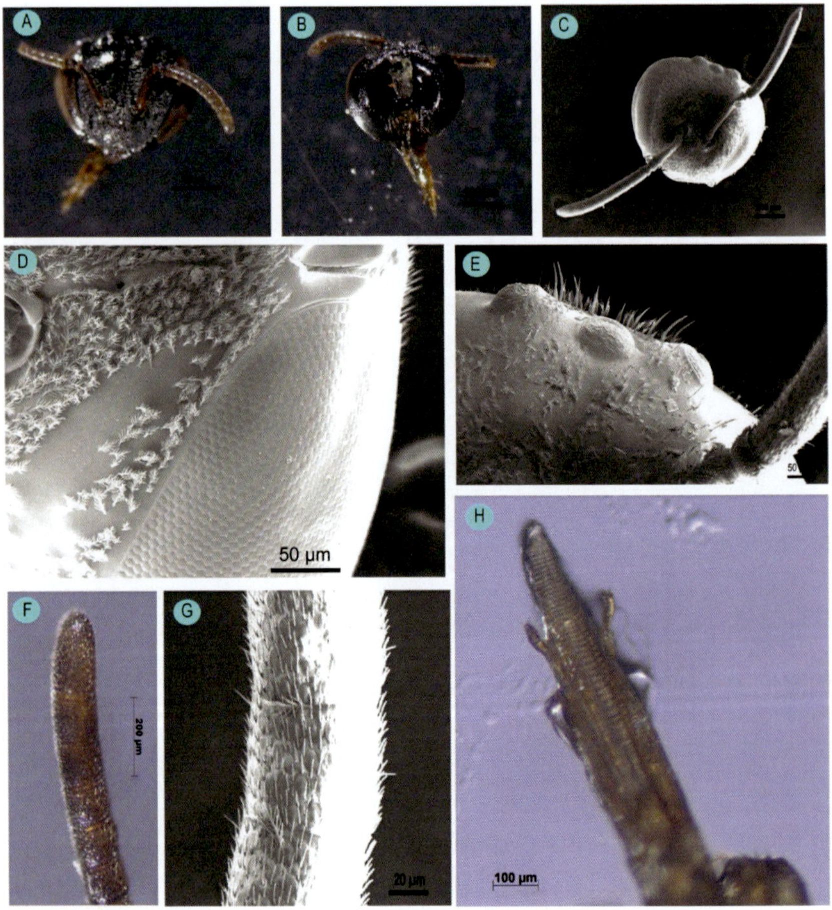

Figure 3.2. External morphology of different parts of the head. (A–C) head, (D) compound eye, (E) ocelli, (F–G) antenna, and (H) proboscis.

B. Thorax

The thorax is blackish, about 1.0 mm in length, and 1.2 mm in width. The dorsal surface bears many unbranched hairs about 80-100 µm long (Figure 2.3). The dorsal side bears two pairs of wings, and the ventral side has three pairs of legs.

Figure 2.3. Thorax showing hairs.

Wings

Fore wings: about 2.5 mm in length and 1.1 mm in width at the distal broadest area and 300 µm. All over the surface bear minute (~20 µm) hairs (Figure 2.4).

Hind wings: about 1400 µm in length and 550 µm in width. All over the surface bear minute (~20 µm) hairs. The number of hamuli on the coastal margin is five.

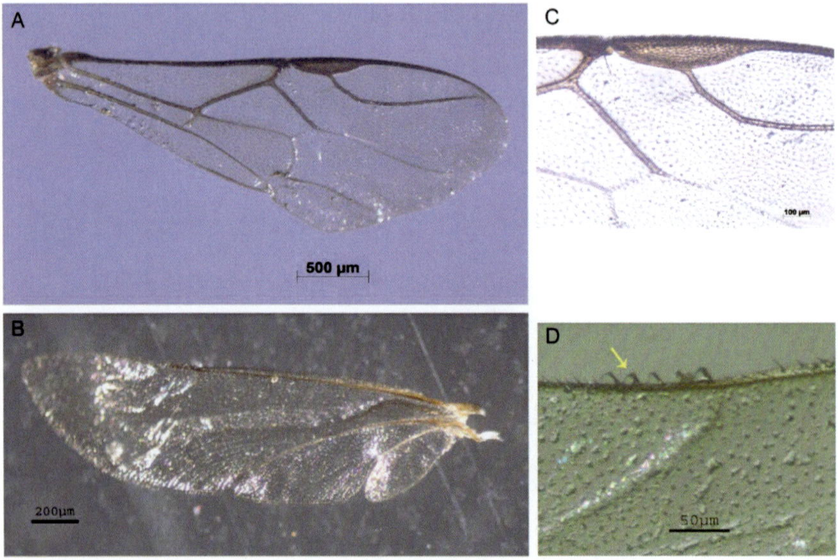

Figure 2.4. Wings of a worker bee. (A) Forewing, (B) hind wing, (C) a portion of the forewing, and (D) a portion of the hind wing showing the hamuli.

Legs

Forelegs: coxa is about 200 µm and bears hairs; trochanter is about 225 µm in length and 100 µm in width, deeply haired; femur is about 500 µm in length and 100 µm in width and bears hairs; tibia is about 400 µm in length and 80 µm in width, bear many hairs, have velum (a fan-like appendage) with 110 µm in length and 55 µm in width (Figure 2.5); tarsi are deeply haired and having antennae cleaner (a notch corresponding to the fan-like appendage); pretarsus with pointed claws.

Middle legs: These are almost similar to the forelegs but do not have fan-like appendages in the tibia and a notch in the tarsi.

Hind legs: coxa is about 250 µm and bears hairs; trochanter is about 250 µm in length and 200 µm in width, deeply haired; femur is about 550 µm in length and 100 µm in width and bears hairs; tibia is about 1000 µm in length, width varies from 700 µm at the lower region and 50 µm in the narrow upper part, bear many hairs, the posterior edge has many-branched hairs about 160 µm in length; tarsi are deeply haired, first segment flat and broad and later segments are narrow; pretarsus with pointed claws.

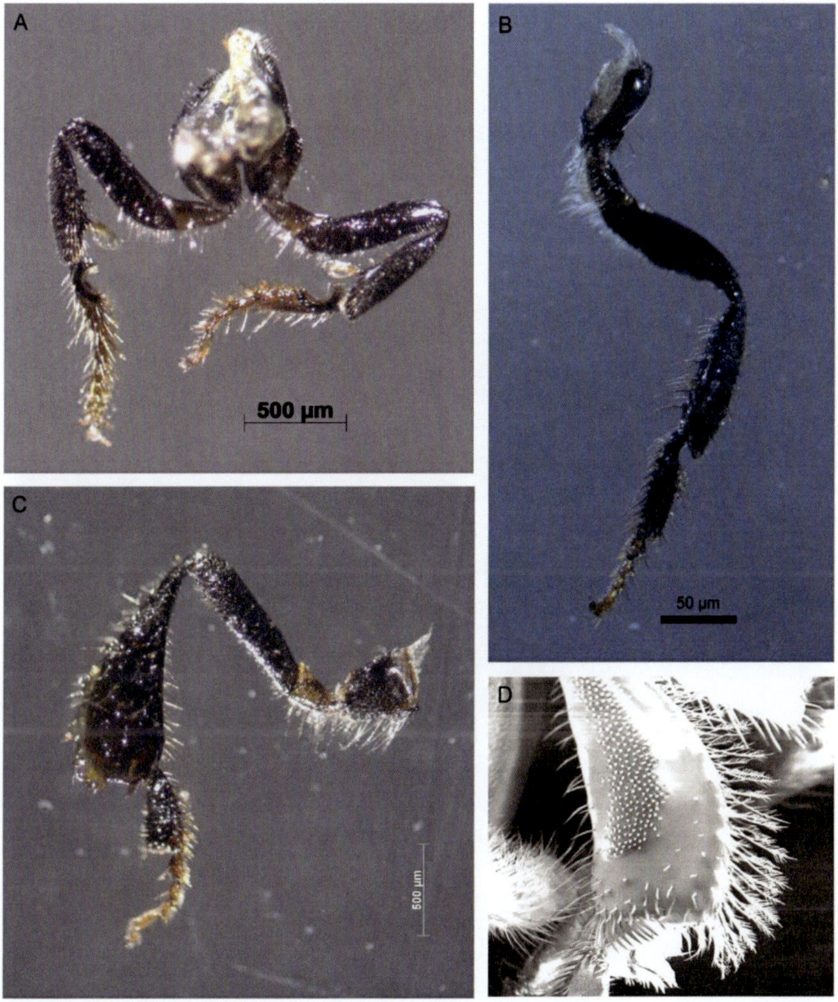

Figure 2.5. Legs of a worker bee. (A) Forelegs, (B) middle leg, (C) hind leg, and (D) corbicula.

C. Abdomen

The abdomen is about 1200 μm in length and 900 μm in width. The distal end is nearly blunt. The abdomen part is amber in colour, with a blackish zone on the dorsal surface at the distal end. Dorsal scales overlap each other and are sparsely haired (Figure 2.6). Ventral scales are also overlapping and much-haired (hairs are unbranched and about 150 μm long).

Figure 2.6. SEM image of the abdomen of a stingless bee.

External morphology of a queen

The body size of the queen is larger than that of the workers and about 7.6 mm long. The head parts are similar to that of a worker bee. The queen thorax is identical to the worker, but the hind legs do not have pollen baskets (Figure 2.7). The abdomen is longer, somewhat taper, and pointed (Figure 2.7).

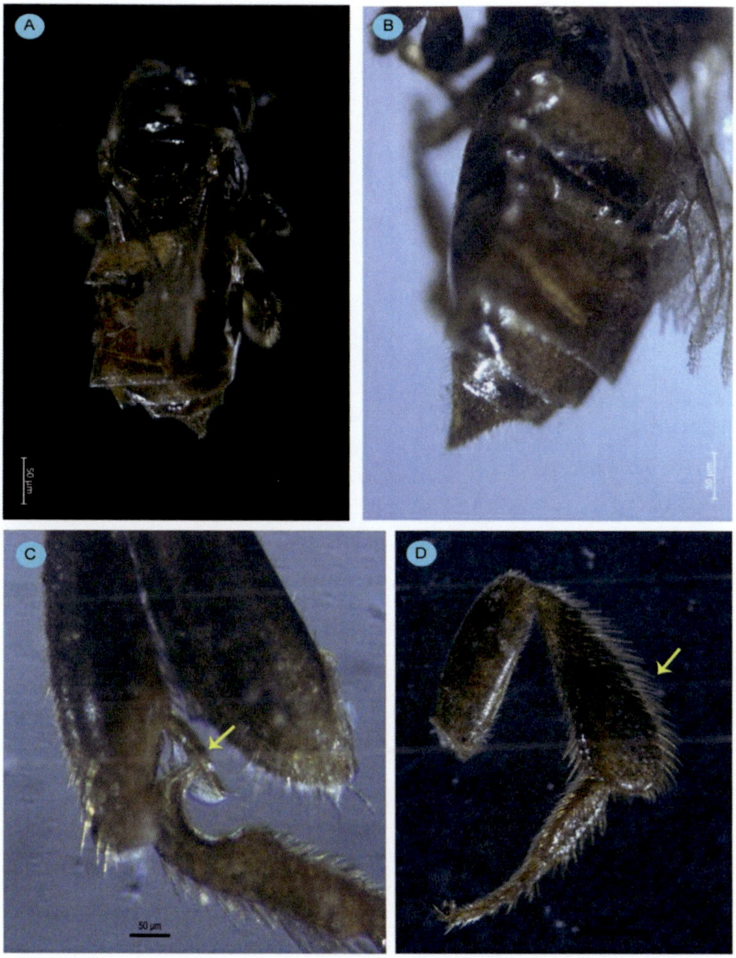

Figure 2.7. (A) A queen stingless bee, (B) abdomen, (C) a part of foreleg showing velum, and (D) a part of hind leg showing ill-developed corbicula.

External morphology of a drone

The drones' bodies are similar in size to the workers' and about 5.1 mm long. Their head parts are similar to those of a worker bee. The drone's thorax is identical to the workers, but the hind legs do not have well-developed pollen baskets. The abdomen is shorter, somewhat blunt, and darker than the workers.

CHAPTER 3

NESTING BIOLOGY OF INDIAN STINGLESS BEES

Nesting substrate

Most researchers use visual-based methods to investigate the nesting substrates for stingless bees. The Indian stingless bees (*Tetragonula iridipennis*) are cavity-nesting species. Their substrate selection may influenced by available empty spaces. They use cavities of tree trunks, building walls, rock crevices, and ridges of agricultural fields (Layek and Karmakar 2018). Other than these, they also colonise inside electric boxes, pipes, and termitariums. Sheltering trees were diverse like *Albizia lebbeck*, *Borassus flabellifer*, *Butea monosperma*, *Ficus benghalensis*, *Ficus religiosa*, *Holoptelea integrifolia*, *Madhuca longifolia* and *Phoenix sylvestris*. Their nested walls are made of mud, mud and stones, mud and bricks, and cement and bricks (Table 3.1; Figure 3.1).

Table 3.1. Nesting substrate for Indian stingless bee.

Nesting substrate	Frequency of occurrence	Source
▪ Building wall		
Made up of mud	+++	Layek and Karmakar 2018; Saaivignesh et al. 2023
Made up of mud and brick	+	Layek and Karmakar 2018
Made up of mud and stone	+++	Layek and Karmakar 2018
Made up of cement and brick	+	Layek and Karmakar 2018
▪ Tree trunk		
Albizia lebbeck	+	Layek and Karmakar 2018
Borassus flabellifer	+	Layek and Karmakar 2018
Butea monosperma	+++	Layek and Karmakar 2018
Ficus benghalensis	++	Layek and Karmakar 2018
Ficus religiosa	++	Layek and Karmakar 2018
Holoptelea integrifolia	+	Layek and Karmakar 2018
Modhuca longifolia	+	Layek and Karmakar 2018
Morus sp.	+++	Choudhary et al. 2021
Phoenix sylvestris	+	Layek and Karmakar 2018
Tectona grandis	+	Own observation
▪ Rock crevices	+	Own observation
▪ Ridge of agricultural field	+	Layek and Karmakar 2018
▪ Pipes		
Iron pipes	+	Saaivignesh et al. 2023
Electrical pipes	+	Nayak Pavithra et al. 2013; Saaivignesh et al. 2023
Water pipes	+	Nayak Pavithra et al. 2013
▪ Electric box	+	Own observation

Note: +++ very frequent, ++ moderate frequent, + less frequent

Figure 3.1. Wild colonies of Indian stingless bees. (A) Mud wall, (B) cement wall, (C) tree trunk of *Butea monosperma*, and (D) tree trunk of *Ficus benghalensis*.

Nest aggregation

The number of stingless bee colonies per site may vary from one to many. However, most sites have a single bee colony, followed by two, three, and more. Higher aggregation can be found only in a few sites. Colony aggregation is more common in walls than other nesting substrates. In the case of aggregated colonies, the minimum distance between the two nearest colonies reported by Choudhary et al. (2021) is about 12.5 cm. However, the very close (2–5 cm distance) situation of nest entrances of the stingless bee colonies is also noticeable (Figure 3.2). Aggregation of colonies may be due to the rapid division of parental colonies into sister colonies, scarcity of nesting substrate, and availability of food resources.

Figure 3.2. Two closely situated nest entrances of Indian stingless bees.

Nest height from the ground level

The heights of the nests from ground level may reach beyond 5 m, with most of the heights being 2–3 m in West Bengal (Layek and Karmakar 2018). Sometimes, stingless bees construct nests very close to ground level (Figure 3.3), especially at sites remote from human settlement. Choudhary et al. (2021) documented that the minimum distance of the nest from the ground in the North-Western plains (Ludhiana) is about 0.75 m. Multiple factors may govern their nesting height, such as flight range, availability of cavities, surrounding vegetation, and disturbance. Stingless bees prefer to construct nests comparatively at lower elevations than wild honeybees (Layek et al. 2023c). The cavity-nesting colonies of stingless bees are more protected than open-nesting colonies of honeybees, even at lower elevations.

Figure 3.3. A wild colony of Indian stingless bees situated very close to the ground level.

Nest orientation

The orientation of the entrance for wild stingless bee colonies can be divided into eight levels: north, south, east, west, northeast, southeast, northwest, and southwest. The different orientations of the nest entrance can be recorded using a GPS inbuilt service and a regular orienteering compass. The nest entrances are oriented in all directions: east, west, north, south, northeast, southwest, northwest, and southeast. Frequencies of entrance orientation vary from region to region. In West Bengal, most surveyed colonies have entrances in the west and north directions (Layek and Karmakar 2018). In Karnataka, higher percentages of nest entrances are recorded in north and north-east directions (Nayak Pavithra et al. 2013). In the north-western plains of India, most colonies preferred making a nest entrance facing east and west (Choudhary et al. 2021).

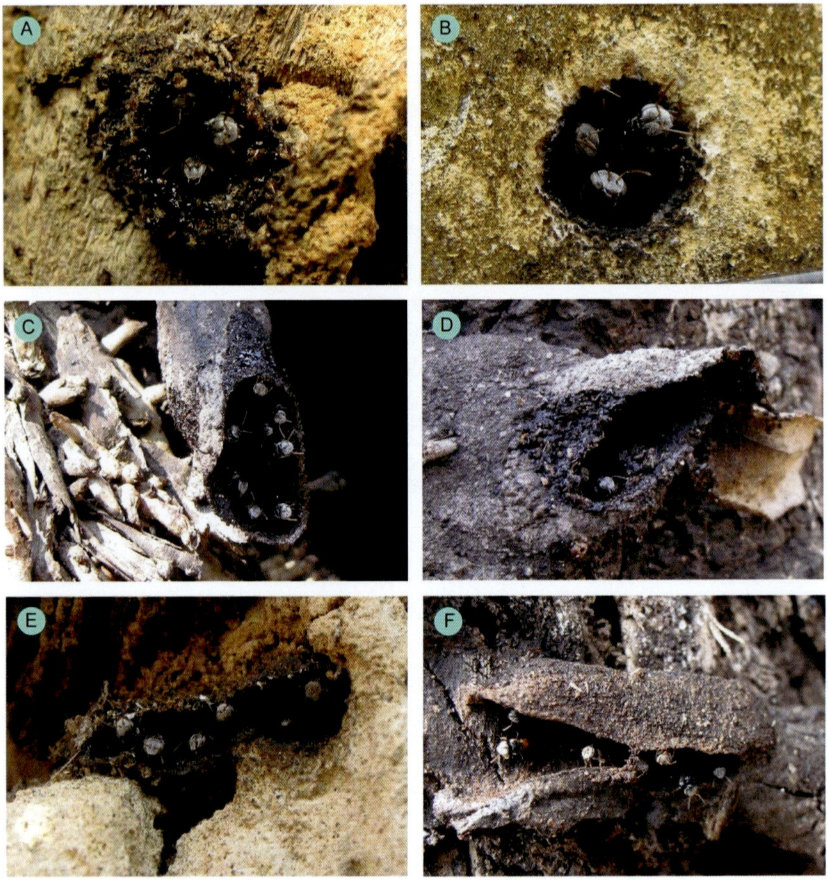

Figure 3.4. Different shapes of nest entrance. (A–B) Circular, (C–D) oval, (E) slit-like, and (F) irregular.

Nest structure

Characteristics of nest entrances can be used to identify species of stingless bees. A few colonies (especially those in a tree trunk) have an external tunnel of 15–78 mm long. The opening of nest entrances is varied in shape, like circular, oval, slit, and irregular (Figure 3.4). The entrance size may be small (about 5×4 mm^2), medium, or large (20×11 mm^2). The colour of the nest entrance may merge with the substratum or sometimes be darker. Nest entrances (or sometimes outside of the external tunnel) had additional deposition of resins, stones, and mud particles (Figure 3.5).

Figure 3.5. Deposition of leaf pieces outside of the external tunnel.

Wild cavity-nesting colonies have an internal tunnel length of about 120.7 ± 35.08 mm (ranges between 84–187 mm). The cavities' dimensions (i.e., length, width, and height) are 204.3 ± 92.45 mm, 159 ± 17.59 mm, and 159 ± 17.59 mm, respectively. Brood cells are located in the centre, while pollen and honey pots are located in the periphery of the cavity. Sometimes, brood cells may develop on one side of the nest, and food pots surround them (Figure 3.6). Besides these components, several involucre and strands support the vital elements, especially brood cells (Figure 3.7). Honey and pollen pots are clumped at many patches in different parts of the nest and may have direct contact with the wall on the inner side. In contrast, brood cells are in single patches within the nest and generally do not touch the side wall directly. Accessory structures mainly involucres are hold them. Brood cells were light brown, oval-shaped, with an average length of 3.45 ± 0.19 mm and width of 2.48 ± 0.26 mm. Queen cells are slightly larger than the cells of workers. Honey and pollen pots are dark brown, oval to polygonal in shape. The length and width of pollen pots are 7.50 ± 0.97 mm and 5.79 ± 0.89 mm, and honey pots are 8.02 ± 0.46 mm and 6.87 ± 0.35 mm. The density (number of cells or pots per cubic inch) of brood cells is 51.9 ± 10.21, while 16.6 ± 2.14 is for pollen pots and 9.55 ± 1.54 for honey pots. The amount of storage pots and brood cells can vary colony-wise and season-wise. Generally, older colonies have more storage pots and brood cells than younger ones. These components also increase in the productive season (when floral resources are more available). At this time, a colony may have several queen cells within the nest. The higher amounts of brood and queen cells indicate good colony growth and assume a high potential for meliponiculture within the region.

Figure 3.6. Nest of Indian stingless bee. (A) Brood cells, (B) honey pots, and (C) pollen pots.

Figure 3.7. Nest showing (A) brood cells for worker, (B) queen cell, (C) strand, and (D) involucres.

CHAPTER 4

COLONY ORGANISATION AND LIFE CYCLE OF INDIAN STINGLESS BEES

Colony organisation

Stingless and honey bees exhibit a transparent caste system with distinct roles assigned to specific groups of individuals. Generally, each colony has several hundreds to thousands of workers, one queen, and several drones (Figure 4.1).

Figure 4.1. (A) Colony of stingless bees, (B) worker bee, (C) queen, and (D) drone. Scale bars = 5 mm.

A. Worker

- The workers make up the bulk of the population of the colony.
- They are imperfectly developed females, morphologically different from the fertile queen in colour and size. The claws of the hind legs of the workers are curved like spinners.
- They perform most of the colony's work except mating and egg-laying. They build the nest, protect the colony, collect resources, and rear broods.
- Workers have an average life span of a few months, depending upon the species. The life span of *Tetragonula iridipennis* workers is about 160 days.

B. Queen

- The queen is the largest caste, identified by its large size, big thorax, and distended abdomen, not covered with wings. The claws of the hind legs are spread across vertically to pre-tarsus.
- Only one queen is found in a colony, but few queen cells can be seen during productive seasons when pollen and nectar flow is high.
- She leaves the nest in only three cases- orientation flight, nuptial flight, and issues a swarm.
- The mating behaviour varies in species, from one to multiple mating. The queen returns to the nest after mating.
- Her primary function is to lay eggs. The queen lays fertilised and unfertilised eggs, ranging between 10–100 depending upon food reserves.
- The queen in some species can live up to 3–7 years.

C. Drone

- They are male members of the colony. The claws of the males are bifurcated for better attachment during mating.
- Drones are typically larger, heavier than female worker bees, and slightly darker.
- They have large, bulging compound eyes, which aid in locating and pursuing queens during mating flights.

- The number of drones within a colony varies with bee species and resource availability. In *Tetragonula iridipennis*, several drones (up to 50) can be noticeable during productive seasons.
- The primary role of drones is to mate with virgin queens from their own or other colonies. This mating behaviour is essential for the colony's reproduction and genetic diversity.

The life cycle of stingless bees

Bees are holometabolous like other insects. The developmental stages of stingless bees include eggs, larvae, pupae, and adults. Their life cycle is similar to other bees but has some unique characteristics.

1. **Colony establishment**: The life cycle of stingless bees begins when a mated queen establishes a new colony. Scout workers search for a suitable nesting site near the mother nest. Upon selection, workers clean and transport food reserves and hive material to the new site, constructing the nest. Once complete, additional workers and a virgin queen migrate from the old nest to the new one.
2. **Egg laying**: The queen lays eggs in the cells. Fertilised eggs become female workers, while unfertilised eggs become male drones. The queen controls the sex of the offspring by fertilising or not fertilising the eggs.
3. **Larval stage**: The eggs hatch into larvae, which feed on the stored food (a mixture of pollen and nectar) within the cells. The larvae go through several moults as they grow, and eventually, they pupate within their cells.
4. **Pupal stage**: The transformation from larvae to adult bees occurs inside the pupal cocoon. During this stage, the body structures and organs develop fully.
5. **Emergence**: Once development is complete, the new adult bees chew their way out of their cells. Initially, they may remain within the nest, taking on tasks like cleaning cells, feeding larvae, or guarding the entrance. As they mature, they take on more complex roles within the colony.
6. **Worker activities**: Stingless bee colonies have a highly organised social structure, with workers performing tasks such as foraging for food, caring for the queen and larvae, building and repairing the nest, and defending the colony.

7. **Reproduction**: At certain times, usually triggered by environmental factors like the availability of resources or changes in weather, a colony may produce new queens and male drones. These reproductive individuals leave the mother colony, and drones mate with virgin queens, facilitating reproduction and genetic diversity.

Cell construction, provisioning, oviposition, and developmental period are essential links between castes (Table 4.1). The queen plays a vital role in regulating the functions of workers, such as cell construction and the discharge of larval food. In some species (e.g., *Lepidotrigona ventralis*: Roopa et al. 2015; *Tetragonula iridipennis*: authors own observation; *Trigona nigrapurpurea*: Wunderink and Sommeijer 1996), cell construction ceased and colony diminished day by day in the absence of a queen. The brood cells of different stages have different colours. The newly constructed cells become brownish and later become straw-coloured with the advancement of age. A single worker can construct one cell in 2–2.5 hrs. After completion, 6–8 attendant workers fill each cell with honey and pollen food mixture. Then, the queen moves around such filled cells by rhythmically fluttering her wings. Later, she deposits an egg within the cell in a vertical or slanting position. Then, workers close the cell with cerumen and finish it within 2–3 min.

Table 4.1. Development period (in days) of different castes of *Tetragonula iridipennis*.

Caste	Egg	Larval stage	Pupal stage	Adult
Worker	5–6	11–13	33–35	80–87
Queen	5–6	16–18	25–26	730–1825
Drone	5–6	19–20	28–29	30–120

Eggs are cylindrical, smooth, and transparent (Figure 4.2). Its length ranges from 1.20–1.37 mm and width from 0.40–0.48 mm. The egg hatches in 5–6 days. The larvae are creamy white, apodous, and 'C' shaped. The length and width of larvae increase with the advancement of age. The larval duration ranged between 11–20 days (depending on the castes). Stingless bees have no physical contact between larvae and adult bees since the cell is closed after egg laying. The pupal cells appear transparent as the brood food is exhausted. The pupae are 4.50–4.75 mm long and 2.90–3.10 mm wide. The pupal period for workers took 33–35 days. The newly emerged workers are light brown; young adults start grooming their bodies and turn darker after eight days. Young workers start nest functions as cleaner bees. After 20–25 days, they serve as guard bees. However, gourd bees of *Tetragonula iridipennis* can collect food staff available close to the nest entrance (author's observation). Generally, after 35–40 days, workers start to forage.

Figure 4.2. (A–B) eggs, (C–D) larvae, (E–H) pupae, (I–J) newly emerged adult bees, and (K) an aged adult bee.

The adult workers' life span is about 80–87 days. Generally, the queen's lifespan is longer than that of worker bees and drones due to her reproductive function and role in colony maintenance. In optimal conditions, a *Tetragonula iridipennis* queen can live for several years, potentially ranging from 2 to 5 years or even longer. However, this lifespan may be shorter if the queen faces stressors such as disease, predation, or inadequate nutrition. Additionally, queens may be replaced if they become less productive or die prematurely, a process known as supersedure. On average, male drones of *Tetragonula iridipennis* typically live for a few weeks to a couple of months. Their primary purpose is to mate with virgin queens, and once this task is completed, they do not contribute significantly to colony maintenance or other activities. After mating, many drones die or are expelled from the colony due to limited resources and space. Environmental factors, such as temperature, humidity, availability of food sources, colony health, and genetics, may also influence the lifespan of male drones. Overall, their lifespan is relatively short-lived compared to the queen and worker bees in the colony.

Stingless bee colonies, like all bee colonies, have a lifespan. Over time, environmental factors, diseases, and other stressors can lead to a colony's decline. However, new colonies can be established by swarming or introducing a new queen to the nest. This cycle continues, with colonies persisting and reproducing as long as environmental conditions remain suitable and resources are available. The life cycle of stingless bees is vital for their survival and the pollination services they provide to their ecosystems.

CHAPTER 5

TASK ALLOCATION OF INDIAN STINGLESS BEES

Different types of tasks

Stingless bees are highly eusocial; many individuals live in a perennial colony. Individuals perform many tasks for the sustenance of a colony, such as nest construction, forage, reproduction, brood care and nursing, nest cleaning, guarding, thermo-regulation, etc. Reproduction is carried out by queens and drones, workers (sterile females) not directly involved in reproduction. Workers perform other tasks. Some major tasks performed by the workers are given below (Figure 5.1):

1. **Construction:** Workers continuously contract brood cells and food pots (nectar and pollen pots) to reproduce and sustain the colony. Young adults are mainly involved in this construction (Figure 5.1A).
2. **Nursing:** The queen and drones require nursing, which workers provide. Workers also offer foodstuffs to larvae before sealing the brood cells.
3. **Cleaning:** The workers who clean the nests are called cleaners. They remove debris outside the colony, exclude cerumen pieces of brood cells after adult emergence, expel deformed larvae and newborn adults, and remove dead bees from the nest (Figure 5.1C-E).
4. **Forage:** Workers collect raw nesting materials like latex, resin, small plant fragments, and sometimes clay particles (Figure 5.1). Workers regularly collect vast amounts of nectar, pollen, and other sugary liquids for colony growth. Few workers who collect only one type of resource on a foraging trip are called specialised foragers (e.g., specialised nectar forager, specialised pollen forager, specialised latex/resin forager). Some other foragers collect two types of resources (nectar and pollen or nectar and latex/resin) on a foraging trip, called mixed foragers (Layek et al. 2022c). This tasks-oriented grouping system is not strictly reserved. A few foragers can change their allocated tasks to other resource-collecting tasks (Layek et al. 2024). Some workers collect water to maintain the nest's

interior temperature and moisture content of honey. The water foragers' proportions fluctuated according to season, with more water foragers during the hot summer.

5. **Guarding:** Some bees guard the colony from predators. Their habitat is at the nest entrance during the daytime. Sometimes, guard bees may be involved in resource collection close to the nest (authors' observation).

Figure 5.1. Different tasks of worker bees. (A) Young bee constructing nest components, (B) gourd bees at the nest entrance, (C–E) cleaner bees removing debris, a deformed larva, and a dead bee, respectively, and (F–I) foragers collecting resin, nectar, pollen, and water, respectively.

Factors influencing task allocation

Many factors, such as age, physiology, environmental conditions, and colony needs, influence task allocation in stingless bee colonies.

1. Age polyethism

Age polyethism is a common phenomenon in eusocial insects, where individuals' roles change as they age. In stingless bees, younger workers typically perform tasks within the nest, while older workers engage in foraging outside the nest.

1. **Young workers (1–10 days):**
   Brood care: Feeding and grooming larvae, cleaning brood cells, and sometimes even capping brood cells.
   Nursing: Producing and distributing food to larvae and other nestmates.
   Hive maintenance: Cleaning the hive, building and repairing nest structures, and processing food brought back by foragers.

2. **Middle-aged workers (10–20 days):**
   Guarding: Protecting the nest entrance from intruders such as predators or bees from other colonies.
   Ventilation: Regulating the temperature and humidity within the hive through wing fanning.

3. **Older workers (20+ days):**
   Foraging: Collecting nectar, pollen, resin, and water. Foragers are critical for sustaining the colony as they gather essential resources.
   Scout bees: Searching for new food sources and potential nesting sites.

Advantages of age polyethism in eusocial stingless bees

Age polyethism behaviour optimises the efficiency and survival of the colony. Here's a breakdown of why age polyethism occurs in these colonies:

- **Task specialisation:** Different tasks in a stingless bee colony require different skills. Younger bees perform safer tasks requiring less experience, such as brood care and cleaning cells. As the bees age, they take on more complex and risky tasks, like foraging outside the hive. This progression ensures bees gain expertise and physical resilience before attempting more perilous tasks.

- **Efficiency:** Age polyethism ensures bees perform tasks aligning with their physiological and behavioural development stages. Younger bees with greater levels of juvenile hormone are better suited for nursing, whereas older bees with lower hormone levels are better suited foraging.
- **Risk management:** By having older bees take on the more perilous work of foraging, the colony reduces the risk to its younger members. This ensures that even if many older bees are lost while foraging, the colony will continue to thrive with a strong cohort of younger bees.
- **Colony health and productivity:** Different tasks necessitate different talents and physical qualities. Younger bees' glands are more active in making brood food, whereas adult bees' flight muscles and navigation abilities are better developed for foraging. This specialisation promotes colony health and productivity.
- **Communication and coordination:** Age polyethism promotes good communication and coordination among the colonies. Older forager bees perform the waggle dance to transmit the location of food sources to their peers, allowing for more efficient resource acquisition.
- **Lifespan maximisation:** By organising tasks according to age group, the stingless bee colony maximises the lifespan and utility of each worker bee. This strategic division of labour helps ensure the colony's long-term sustainability and success.

2. Role of physiology in task allocation

Certain physiological factors influence the task allocation of stingless bees. For instance, juvenile hormone (JH) and vitellogenin levels can affect behavioural development and task specialisation.

1. **Hormonal regulation:**
   In stingless bee colonies, the allocation of tasks is a multifaceted process influenced by various factors, including hormonal signals. Two primary hormones, juvenile hormone (JH) and ecdysteroids, are essential in this regulatory mechanism. Juvenile hormone, synthesised in the corpora allata, significantly impacts the age-related division of labour, referred to as temporal polyethism. As bees mature, their levels of JH rise, prompting a shift from in-hive

responsibilities, such as nursing, to foraging activities outside the hive. Elevated JH levels correlate with the initiation of foraging behaviour, as this hormone facilitates the physiological adaptations required for this energy-intensive task. Ecdysteroids, while primarily recognised for their involvement in insect development, also contribute to task allocation in stingless bees, albeit in a less direct manner than JH. The levels of ecdysteroids vary in response to environmental and social stimuli, influencing the expression of genes associated with metabolism, immunity, and stress responses. These hormones interact with additional factors, including pheromones and social dynamics within the colony, to precisely adjust the behavioural roles of individual bees, thereby ensuring the colony operates effectively as a superorganism. The careful regulation and timing of JH and ecdysteroid levels are vital for sustaining the colony's adaptable workforce, enabling it to respond to fluctuating internal and external conditions.

2. **Gland development:**
The development of glands in stingless bees is closely associated with their allocation of tasks, which is essential for defining their roles within the colony. Notably, the hypopharyngeal glands, wax glands, and ovarian development are critical in this regard. The hypopharyngeal glands, found in the heads of worker bees, produce royal jelly, a nutrient-rich substance utilised for feeding both larvae and the queen. These glands are most prominent in young nurse bees, whose primary responsibility is to nurture the brood. When bees mature and shift to foraging activities, the functionality of these glands decreases, reflecting their changing responsibilities. In a similar vein, the wax glands located on the abdomen are crucial for bees involved in the construction and repair of nest components. These glands are most active in worker bees during the mid-stage of their lifespan when they are responsible for building and maintaining the hive. The secretion of wax is vital for forming the brood cells and storage pots that accommodate brood and store honey and pollen. Although ovarian development is generally inhibited in worker bees, it can be triggered in the absence of a queen, prompting some workers to lay unfertilised eggs that develop into drones. This capacity for ovarian activation highlights the adaptability of reproductive roles within the colony, influenced by the queen's social dynamics and presence. The development and regression of these glands are intricately aligned with the bees' age and the colony's

requirements, ensuring that the appropriate individuals are prepared for the necessary tasks at any moment. This physiological adaptation enhances the colony's efficiency, facilitating a smooth division of labour that responds to internal and external pressures.

3. **Neurophysiology:**
Neurophysiology, especially concerning brain development, is crucial for task allocation among stingless bees, influencing the behavioural specialisation vital for their colonies' success. In both types of bees, significant alterations in brain structure occur as they progress through various life stages and take on different responsibilities within the colony. The mushroom bodies, paired structures in the insect brain, play a key role in this transformation, as they are integral to sensory processing, learning, and memory functions. Young bees, primarily engaged in in-hive activities such as nursing, exhibit less developed mushroom bodies, indicative of the lower cognitive demands associated with these tasks. Conversely, as bees mature and transition to more intricate roles like foraging, their mushroom bodies experience considerable growth and synaptic reorganisation. This development is influenced by age and experiential factors, with heightened foraging activity promoting neural plasticity, thereby improving the bees' navigation, learning, and memory of floral resources. Similar neurophysiological transformations are noted in stingless bees, although the details may differ based on species and ecological context. The neural circuitry also processes sensory information from the environment and pheromonal signals within the hive, allowing bees to respond to the evolving needs of the colony adaptively. For example, neurophysiological mechanisms, likely modulated by hormonal changes and environmental cues, encourage specific individuals to shift to foraging duties when the colony requires additional foragers. These neurodevelopmental processes ensure that bees are adequately prepared physically and cognitively for the demands of their designated tasks, thereby promoting an efficient division of labour. The dynamic nature of brain development in stingless bees illustrates the complex relationship between neurophysiology and social structure, highlighting the adaptability and resilience of these eusocial insects.

4. **Protein mediates regulation**
   Proteins, especially vitellogenin, are essential in managing task distribution among bees, serving as vital intermediaries in the intricate relationship between physiology, behaviour, and social organisation. Vitellogenin, a precursor protein for yolk primarily linked to reproduction in various insects, has been shown to affect labour division within stingless bee colonies. In worker bees, which generally do not reproduce, vitellogenin levels are closely associated with age-related task distribution, known as temporal polyethism. Young nurse bees exhibit high concentrations of vitellogenin, which aids in brood care by supplying necessary nutrients for developing larvae. As these bees mature and shift to foraging roles, vitellogenin levels decline while juvenile hormone levels rise, promoting this behavioural transition. The interplay between vitellogenin and juvenile hormone is vital in determining the timing of a bee's shift from in-hive responsibilities to foraging activities, with vitellogenin also providing a protective function against oxidative stress, which is particularly crucial given the strenuous nature of foraging. Furthermore, vitellogenin affects the lifespan of worker bees, with elevated levels correlating with increased longevity, enabling some bees to continue nursing roles longer if the colony's demands necessitate it. This protein also plays a role in social regulation, as it interacts with pheromones and other signalling molecules to help sustain colony stability and adapt to fluctuating environmental conditions. In summary, vitellogenin acts as a molecular bridge connecting nutrition, reproductive physiology, and behaviour, facilitating the effective distribution of tasks within the colony. This ensures that the workforce is optimally allocated to address the evolving needs of the colony, from brood care to resource acquisition, underscoring the complex biological mechanisms that support social structure in bees.

3. Role of genetics in task allocation

1. **Genetic predisposition:**
   Genetic predisposition is crucial in determining bee task allocation, with specific genes linked to particular roles and colony-level selection enhancing specialisation and efficiency. Studies indicate that specific genes correlate with behaviours essential for the colony's upkeep, including foraging, nursing, and guarding. The expression of these task-specific genes varies according to the bee's

age, environmental factors, and the colony's requirements, resulting in the clear division of labour characteristic of eusocial insects. For instance, genes associated with sensory perception, learning, and navigation are more active in foraging bees, whereas those related to nutrition and immune responses are predominantly expressed in nurse bees. This genetic framework ensures that individual bees are naturally inclined to perform specific functions, thereby boosting the colony's overall productivity and survival.

Additionally, the selection at the colony level significantly impacts the distribution of tasks, as natural selection favours those colonies that effectively allocate responsibilities among their members. Over time, colonies that enhance task distribution through genetic diversity and adaptive mechanisms are more likely to prosper. This selection mechanism can develop behavioural subcastes within the worker group, where specific lineages exhibit a greater propensity for particular tasks. The genetic diversity present within the colony is also essential, as it offers a broader array of behavioural options that can be adjusted in response to environmental and social stimuli. Such diversity enables the colony to adapt to varying circumstances, including changes in resource availability or threats posed by predators and parasites. Therefore, genetic predisposition, influenced by specific task-related genes and overarching colony-level selection, plays a vital role in managing task allocation in bees, facilitating the maintenance of the complex social structure required for their collective success.

2. **Gene expression:**

Gene expression, shaped by epigenetic factors and nutritional inputs, is crucial for managing task distribution among bees, enabling them to adjust their behaviours in response to the evolving requirements of the colony. Epigenetic processes, including DNA methylation and histone modification, influence gene expression without changing the fundamental genetic sequence, allowing bees to display varied behaviours according to their environmental and social circumstances. For example, in honeybees, epigenetic alterations can influence whether a larva matures into a queen or a worker, with nutrition being a significant factor in this determination. The feeding of royal jelly, a nutrient-dense substance provided to prospective queen larvae, initiates specific epigenetic changes that either activate or inhibit essential developmental genes, resulting in the differentiation of queens, who are tasked with reproduction, from workers, who undertake various roles within the colony.

As worker bees mature, epigenetic modifications persistently affect their distribution of tasks. For instance, alterations in methylation patterns may occur in response to the changing demands of the colony, leading a worker to shift from nursing duties to foraging activities. Furthermore, nutritional inputs, such as the consumption of pollen or nectar, play a significant role in regulating these epigenetic alterations, which in turn influence the expression of genes associated with metabolism, immunity, and behaviour. Similar processes are observed in stingless bees and other eusocial species, where the type and amount of food ingested by both developing larvae and adult workers shape their gene expression profiles, ultimately determining their specific roles within the colony.

Moreover, nutritional signals engage with the endocrine system, particularly hormones such as juvenile hormone and vitellogenin, which are crucial in regulating gene expression related to task execution. The interaction between nutrition, epigenetics, and gene expression enables the colony to dynamically assign tasks among its members, ensuring that the appropriate individuals are ready for foraging, nursing, or other responsibilities as required. This regulatory framework is vital for sustaining the colony's efficiency, resilience, and adaptability to environmental challenges, underscoring the complex relationship between gene expression, epigenetic regulation, and bee social behaviour.

3. **Behavioural genetics:**

Behavioural genetics, especially concerning polyandry, genetic variation, kin selection, and altruism, is essential in determining task distribution within bee colonies, enhancing their social intricacy and operational efficiency. Polyandry, characterised by a queen mating with several males, significantly boosts the genetic variation present in the colony. This increased genetic variation is advantageous as it leads to diverse behavioural phenotypes among the worker bees, improving the colony's capacity for effective task allocation. Within a single colony, different genetic lineages may exhibit predispositions toward particular roles, such as foraging, nursing, guarding, or comb construction, influenced by their inherited behavioural characteristics. This division of labour enables the colony to respond more adeptly to fluctuating environmental conditions and resource availability, ensuring that the workers most suited to each specific function carry out all essential tasks.

Kin selection and altruism significantly shape task allocation by promoting behaviours that contribute to the survival and reproductive success of the colony, even at a personal cost to individual workers. Given the high genetic relatedness among workers, they are more inclined to engage in altruistic actions that benefit the colony, such as defending against predators or nurturing the queen's offspring. These altruistic tendencies are supported by kin selection, which maximises an individual's inclusive fitness by propagating shared genes among relatives. This genetic inclination towards altruism ensures that essential tasks for the colony's survival, including brood care and hive upkeep, are prioritised, even when they do not yield direct benefits to the individual worker.

The interaction of polyandry, genetic diversity, and kin selection establishes a robust framework for task allocation, wherein the genetic composition of the colony facilitates a division of labour that enhances overall fitness. This resulting behavioural adaptability enables the colony to prosper in various and evolving environments, as different workers can be assigned to specific tasks according to their genetic traits and the immediate requirements of the colony. This genetic influence on task allocation highlights the evolutionary benefits of social living in bees, where the colony's overall success relies on its members' efficient coordination and collaboration.

4. Environmental factors influencing task allocation

External conditions like the availability of resources and climatic factors can also influence task allocation. For example:

1. **Resource availability**
   The availability of resources, particularly food and water, significantly impacts the allocation of tasks among bees, thereby influencing the behaviour and labour distribution within the colony. When food sources, such as nectar and pollen, are plentiful, the colony can assign a greater number of workers to foraging activities, thereby enhancing the collection and storage of these essential resources. Foragers are sent out to take advantage of these abundant food supplies, and the communication methods employed within the colony, such as the waggle dance performed by honeybees, effectively guide other workers to productive foraging locations. This increase in foraging efforts may also prompt younger bees to

shift their roles more rapidly, moving from in-hive responsibilities like nursing to foraging duties in response to the heightened need for resource acquisition.

In contrast, when food sources become limited, the colony must modify its labour distribution to conserve energy and maximise resource utilisation. During these periods, fewer bees are dispatched for foraging, and those who venture out may have to cover greater distances or exert more effort to find adequate food. This scarcity can delay the transition of younger bees into foraging roles, necessitating their continued presence in the hive to carry out vital maintenance and brood care tasks that are crucial for the colony's survival during challenging times. Furthermore, limited food availability may intensify worker competition, affecting which bees are more actively involved in foraging based on their efficiency or prior experience.

Water availability is paramount, particularly in regulating the colony's temperature and humidity, which are essential for developing the brood. In excessively hot or dry conditions, more worker bees may be designated for water collection, ensuring the hive remains cool and that the larvae receive adequate hydration. This responsibility becomes particularly vital during extreme weather events when it is crucial to maintain the hive's internal environment. The reassignment of bees to water collection can also decrease the workforce available for other duties, highlighting the adaptive nature of task distribution in response to changing environmental factors.

Resources such as food and water govern the immediate allocation of tasks among the bees. This impacts the long-term configuration of the colony's workforce, enabling it to adjust to varying environmental conditions while preserving its health and productivity.

2. **Seasonal changes**

Seasonal variations play a crucial role in determining the distribution of tasks among bees, prompting alterations in their behaviour and the dynamics of the colony that are vital for their year-round survival. During the spring and summer months, characterised by abundant flowers and warmer temperatures, the colony experiences a surge in activity. The primary objective during these seasons is foraging for nectar and pollen, which are critical for honey production and nurturing the brood. The queen's egg-laying rate increases, resulting in a larger population of worker bees available for various responsibilities. Younger bees typically take on the role of nurse bees, caring for the brood, while older workers focus on foraging. The foraging contingent expands to take advantage of the plentiful resources, with many bees leaving the hive to collect food. Furthermore, bees participate in hive maintenance activities, such as building comb, as the heightened nectar flow requires additional storage capacity. The overall activity level within the colony is elevated, characterised by a well-organized division of labour that optimises resource gathering in preparation for the approaching winter.

The colony's focus shifts as autumn arrives and floral resources become less plentiful. The queen decreases her egg-laying, resulting in a gradual reduction of the bee population. Task allocation becomes increasingly centred on winter preparations, with bees prioritising consolidating and capping honey stores to ensure sufficient food supplies for the colder months ahead. While foraging persists, fewer bees are assigned to this task as resources dwindle. Some workers may be reassigned to defensive roles, safeguarding the hive entrance from potential intruders seeking to pilfer the precious honey reserves. Additionally, bees may adopt behaviours aimed at conserving energy, such as clustering within the hive to maintain warmth.

In the winter months, the activity level of the colony experiences a marked decline, with task distribution primarily focused on survival rather than the accumulation of resources. The bees gather closely around the queen to retain warmth, and only a few workers venture outside unless the weather is exceptionally mild. Foraging activities nearly halt, compelling the bees to depend on the honey reserves collected during the warmer seasons. The emphasis transitions to regulating the hive's temperature and safeguarding the queen and the essential workers. Bees that participated in foraging during the summer may now take on responsibilities related to temperature management, such as generating heat through muscular activity or providing insulation for the cluster. This seasonal reallocation of tasks is vital for the colony's survival. It enables the bees to optimise their energy expenditure and resource management throughout the year, thereby ensuring the colony's persistence across different seasons.

3. **Colony demographics**

The demographics of a colony, particularly the brood's existence and the queen's health status, along with her pheromonal signals, are pivotal elements that affect the distribution of tasks among bees, thereby determining how labour is organised to fulfil the colony's needs. The presence of brood, encompassing eggs, larvae, and pupae, is a significant motivator for task allocation within the hive. When brood is present, many worker bees are designated nursing responsibilities, nourishing the developing larvae with royal jelly, pollen, and nectar. This necessity for brood care is a crucial factor influencing the roles assumed by younger bees, as they are both biologically and behaviorally suited for tasks such as feeding and cleaning the brood. The level of nursing activity is directly proportional to the quantity of brood present in the colony; a larger brood necessitates a more significant number of nurses, resulting in an increased percentage of workers engaged in these activities.

The health of the queen and the pheromones she secretes are also essential in managing task allocation. A healthy queen releases potent pheromones disseminated throughout the colony by worker bees, indicating her presence and reproductive condition. These pheromones play a vital role in inhibiting the reproductive functions of worker bees, ensuring their focus remains on non-reproductive activities such as foraging, nursing, and maintaining the hive. Additionally, the queen's pheromones affect the behavioural development of workers, regulating the timing of their shift from in-hive responsibilities to foraging tasks. Should the queen's health deteriorate or her pheromone output decrease, it may disrupt task allocation, with some workers potentially activating their ovaries to lay eggs, undermining the colony's social structure and operational efficiency.

Furthermore, the pheromones released by the queen play a crucial role in shaping the colony's unified reaction to external threats and environmental shifts. For instance, a decline in the strength of the queen's pheromonal signals may prompt the worker bees to initiate preparations for the potential replacement of the queen, thereby reallocating resources and labour towards nurturing a new queen. This transition can result in temporary alterations in task distribution, with an increased number of workers dedicating their efforts to the care of the specialised queen larvae. The complex relationship between the presence of the brood, the health of the queen, and her pheromonal communications ensures that the colony's workforce is effectively organised to address the evolving requirements of the hive, thereby preserving both harmony and productivity within the colony.

4. **Colony stressors**

Stressors affecting bee colonies, including predators, parasites, and pathogens, significantly influence task allocation, prompting behaviour and labour distribution alterations to address threats and sustain colony health. In instances of predation, such as assaults by wasps or hornets, there is a notable shift in task allocation towards defensive measures. Worker bees are redirected from their typical responsibilities of foraging and caring for the brood to focus on guarding the hive and repelling intruders. This defensive response is essential for safeguarding the colony's resources and reproductive members, often necessitating heightened vigilance at the hive entrance and coordinated group efforts to confront predators.

Parasites like Varroa mites and pathogens, which encompass various bacteria, viruses, and fungi, further disrupt task allocation by collectively compromising the health of individual bees and the colony. Infestations or infections may result in a greater emphasis on cleaning and grooming behaviours as bees strive to eliminate parasites and infected individuals from the hive. Ailing or weakened bees might be segregated from the healthier workers. In extreme situations, the colony may reassign workers to undertake additional hygienic tasks, such as removing diseased broods or waste management, to curb the spread of pathogens. The presence of these pathogens can also diminish the overall productivity of the colony, leading to a decrease in the number of foragers and an increased dependence on stored resources as the colony's capacity to gather new supplies wanes.

In addition, the colony's reaction to these stressors frequently involves alterations in its reproductive and developmental focus. For example, when faced with elevated levels of disease or parasitic threats, the queen may curtail egg production to limit the number of susceptible offspring. Conversely, the colony might emphasise nurturing new queens if the existing queen's health is at risk or if there is a notable decline in the worker bee population due to stress. This tactical adjustment enables the colony to sustain its operations and reproductive capabilities despite unfavourable circumstances. How a stingless bee colony distributes responsibilities in response to stressors illustrates its flexible and adaptive characteristics, facilitating its ability to uphold stability and resilience amid challenges that jeopardise its survival and productivity.

5. **Climatic conditions**
Climatic factors, such as temperature, humidity, and other weather-related elements, are crucial in determining the task distribution among bees, thereby influencing their behaviour and operational efficiency. Temperature is a critical factor for driving activity within the colony, influencing the bees' capacity to undertake diverse tasks. Bees exhibit heightened activity levels in warmer conditions, allowing them to engage more frequently and effectively in foraging, nectar collection, and hive maintenance. The abundance of floral resources during the warmer months increases foraging efforts, as bees are sent out to collect nectar and pollen to meet the colony's requirements. In contrast, during periods of extreme heat, bees may prioritise regulating the hive's temperature by employing behaviours such as evaporative cooling, where they fan their wings to enhance air circulation and lower internal temperatures. This cooling mechanism is vital for maintaining the optimal temperature range essential for brood development and the overall health of the hive.
As temperatures drop, particularly in autumn and winter, the allocation of tasks undergoes a significant transformation. Foraging activities are diminished or ceased as bees depend on their stored honey and pollen to sustain the colony throughout the colder months. The bees cluster around the queen and the brood to conserve heat, with task distribution concentrating on preserving the warmth of this cluster and safeguarding the hive from cold and moisture. During these colder periods, fewer worker bees are available for external activities, leading to a shift in focus towards internal hive maintenance and temperature control. Furthermore, elevated humidity levels can affect hive conditions, potentially increasing fungal and bacterial proliferation. Bees may modify their behaviours to regulate humidity within the hive by enhancing ventilation to lower moisture levels or growing hive grooming surfaces to mitigate mould and disease.

5. Communication and coordination influencing task allocation

Communication and coordination are essential for efficient task allocation in eusocial bee colonies. These processes ensure bees work harmoniously and adaptively to meet the colony's needs. Here are the key ways in which communication and coordination facilitate task allocation:

1. **Trophallaxis**
   Trophallaxis, the mechanism through which food or fluids are exchanged among colony members, is vital in determining bee task distribution, especially in social species such as honeybees. This complex behaviour facilitates the exchange of nutrients and the communication of essential information, which is critical for the colony's efficiency and well-being. Bees can share crucial nutrients and energy sources through trophallaxis while transmitting chemical signals regarding the colony's requirements and environment. For instance, when a bee ingests a nutrient-dense substance, it can communicate its nutritional status to others, influencing their actions and responsibilities. This process aids in coordinating activities like foraging, caring for the brood, and maintaining the hive, ensuring that tasks are allocated to the most capable individuals based on their physiological conditions and environmental signals. Additionally, trophallaxis can disseminate alarm signals and pheromones, thereby modifying the colony's reaction to external threats or changes. Consequently, trophallaxis plays a significant role in the division of labour, enhancing the overall efficiency and adaptability of the colony in response to various challenges.

2. **Pheromones**
   Pheromones are essential in managing task distribution within bee colonies, with various types—such as queen pheromones, brood pheromones, and alarm pheromones—each affecting the behaviour and responsibilities of worker bees in unique manners. Queen pheromones are vital for sustaining social unity and reproductive order; they are secreted by the queen and spread throughout the hive, indicating her presence and reproductive condition to the workers. This chemical signal ensures that worker bees concentrate on duties such as caring for the brood and maintaining the hive rather than engaging in reproductive activities. Brood pheromones, released by larvae and pupae, prompt worker bees to prioritise the care and feeding of the brood, thereby facilitating the allocation of resources and attention to the young. These pheromones also indicate the

necessity for specific tasks, including cleaning and safeguarding the brood. Conversely, alarm pheromones react to threats or disturbances, activating the colony's defensive behaviours. When a worker bee identifies an intruder or senses danger, it releases alarm pheromones that initiate a coordinated defensive response among the other bees, ensuring a rapid and organised effort to protect the colony. The interaction of these pheromones allows bees to effectively manage their responsibilities and sustain colony operations by continuously responding to internal and external factors.

3. **Vibration signals**
Vibration signals constitute a vital communication component within bee colonies, significantly affecting task distribution and organisational structure by prompting specific actions among worker bees. These vibrations, generated through various methods such as wing fanning, body shaking, or direct contact, act as essential indicators for synchronising hive operations and managing responsibilities. A prime example is the "waggle dance," a notable vibration-based behaviour exhibited by foraging bees, which not only conveys the location of food sources but impacts the allocation of foraging duties among the workers. Upon returning to the hive and executing the waggle dance, a forager produces vibrations that inform other bees about the direction and distance to the food source, motivating them to embark on foraging expeditions. Likewise, vibrations within the hive can trigger activities related to brood care and maintenance. For instance, worker bees may utilise vibrations to indicate the necessity for increased brood attention or promote larvae cleaning and grooming. Furthermore, vibrations can influence the hive's reaction to environmental shifts or threats; a sudden vibration may alert bees to potential dangers, initiating defensive actions and redistributing tasks to safeguard the colony. The vibration signals offer a dynamic and immediate method for coordinating activities, enabling stingless bees to adapt to their constantly changing environment and effectively assign tasks to sustain colony health and productivity.

4. **Acoustic signals**
Acoustic signals, especially those produced by the queen bee, are crucial for determining task distribution within bee colonies. Queen piping refers to the unique sounds generated by the queen,

particularly during significant events such as mating flights or swarm preparations. These sounds are instrumental in sustaining the social hierarchy and guiding the actions of worker bees. For instance, during the queen's nuptial flights, she produces specific piping sounds that indicate her presence and readiness to mate, facilitating the coordination of drone behaviour and ensuring their involvement in mating processes. Within the hive, the queen's piping is an important marker of her health and reproductive condition. When the queen is healthy and actively laying eggs, her piping reassures the worker bees, reinforcing their commitment to brood care and hive upkeep. In contrast, a decrease or irregularity in the queen's piping may indicate potential health or fertility issues, prompting worker bees to engage in activities such as grooming the queen, preparing for swarming or even initiating the development of a new queen. Furthermore, queen piping can affect the colony's overall reaction to environmental shifts or internal disruptions by influencing worker behaviour and ensuring that essential tasks, such as brood rearing or hive defence, are effectively assigned. The queen piping is a critical acoustic signal that regulates task allocation, enabling the colony to operate cohesively and efficiently in response to various stimuli.

5. Spatial organisation

Spatial organisation, also called task zoning, serves as a critical mechanism that governs the distribution of tasks within stingless bee colonies, thereby ensuring that various activities are effectively allocated across distinct areas of the nest. In these bees, the hive is generally segmented into specialised zones tailored to the specific tasks being executed, with each zone fulfilling a unique function such as brood rearing, food storage, or hive upkeep. This spatial differentiation promotes a more systematic and efficient approach to colony management. For instance, the central region of the nest may be designated for brood cells, where larvae and pupae receive care. At the same time, the outer areas are typically utilised to store nectar, pollen, or honey. The careful arrangement of these zones optimises the use of space and resources, minimises the potential for task overlap, and enhances the overall efficiency of the colony. Furthermore, task zoning regulates hive temperature and humidity by situating temperature-sensitive activities, like brood care, in areas with more stable environmental conditions. Worker bees are often allocated to specific zones according to their roles and the immediate requirements of the colony; for example, bees tasked with brood care

are predominantly found in the central brood area, whereas foragers and food stores operate within the storage zones. This spatial organisation mitigates congestion and ensures that each task is conducted appropriately, fostering a well-coordinated and functional colony structure. The spatial organisation through task zoning is essential for optimising task allocation, boosting colony productivity, and sustaining stingless bee hives' overall health and efficiency.

6. **Social interactions**
   Social interactions are crucial in shaping task allocation among stingless bees, with the ability to switch tasks being a fundamental element of this process. Unlike fixed task assignments, task switching enables these bees to modify their roles following the colony's needs and immediate environmental demands. Interactions such as direct contact, grooming, and trophallaxis promote this adaptability by allowing bees to communicate and react to shifts in the colony's requirements. For example, when a bee engages with others through grooming or feeding, it can detect signals that suggest a change in task priorities, such as an urgent need for additional foragers or increased brood care. This flow of information can lead bees to transition smoothly from one task to another, ensuring that the colony remains agile in response to variations in resource availability, environmental factors, or internal hive necessities. Furthermore, the presence of seasoned bees performing specific tasks can encourage less experienced workers to adopt those roles through observation and interaction. This flexible task switching improves the colony's efficiency by fostering a more dynamic approach to task allocation, allowing bees to continually adjust their roles to enhance colony function and ensure the effective completion of all essential tasks. The capacity to switch tasks based on social interactions enables stingless bees to sustain a well-organized and resilient colony structure, adept at thriving under diverse conditions and quickly adapting to changing situations.

7. **Learning and memory**
   Learning and memory processes are pivotal in determining task allocation among stingless bees, influenced by environmental and social learning mechanisms. Environmental learning refers to the bees' ability to gather information about their surroundings, including the locations of food sources, the most effective foraging

paths, and the conditions within the hive. This form of learning equips bees to navigate their environment adeptly and adjust their roles according to their experiences. For example, forager bees utilise environmental signals to locate and utilise food sources, which is essential for sustaining the colony's food supply. Conversely, social learning transpires when bees observe and replicate the actions of their fellow nestmates. This learning is vital for tasks requiring specific skills or knowledge, such as collecting nectar or caring for the brood. Inexperienced or younger bees typically acquire these skills by observing seasoned individuals who offer guidance and demonstrate proficient techniques. As bees learn and retain these abilities, they can transition between tasks or assume new responsibilities within the colony as required. For instance, a bee that learns effective foraging techniques by watching experienced foragers may eventually adopt this role, thereby enhancing the colony's foraging capacity. Furthermore, memory is essential for enabling bees to remember the locations of food sources, the hive's layout, and the dynamics of social interactions, all of which contribute to optimising task allocation and ensuring efficient task execution. Environmental and social learning allows stingless bees to adapt to evolving circumstances, acquire essential skills, and effectively distribute tasks within their colony, promoting individual and collective success.

8. **Feedback mechanisms**

Feedback mechanisms, especially self-regulation, play a crucial role in determining task distribution among stingless bees, enabling the colony to adaptively modify tasks in response to immediate needs and environmental conditions. Self-regulation in these bees entails ongoing observation and modification of behaviours, guided by feedback from both the internal dynamics of the colony and its members. For example, when bees engage in specific activities such as foraging or caring for the brood, they receive feedback through pheromones, vibrations, and variations in the hive's environment. Bees can adjust their actions if a task is executed ineffectively or if there is a sudden shift in the colony's requirements. For instance, if nectar availability decreases, foraging bees may intensify their foraging activities. In contrast, others may shift their focus to brood care to ensure the proper nurturing of the colony's young. Likewise, when a specific task is either overly or insufficiently represented, feedback mechanisms facilitate the redistribution of functions to achieve a balanced workload. This adaptability is evident in how worker bees modify their roles to the immediate needs of the hive, such as transitioning from brood care to hive maintenance or the reverse. Furthermore, these feedback mechanisms assist in regulating internal hive conditions, including temperature and humidity, by modulating the activities of bees responsible for temperature control or ventilation. Through these self-regulatory mechanisms, stingless bees maintain a flexible and responsive task allocation system that addresses internal and external fluctuations, ensuring optimal colony performance and enhancing overall productivity.

6. Flexibility in task allocation

Stingless bee colonies exhibit a high degree of flexibility in task allocation. Workers can switch roles in response to the colony's needs or changes in environmental conditions.

1. **Behavioural plasticity:**

   Role-switching: Role-switching plasticity plays a vital role in allocating tasks among stingless bees, allowing them to adapt effectively to their colony's changing conditions and demands. This adaptability enables individual bees to shift between various roles depending on their current status, experience, and the colony's needs. For instance, a bee primarily focused on caring for the brood may transition to foraging duties when there is a heightened need for food resources or a shortage of forager bees. This capacity for role transition is supported by the bees' ability to quickly learn new tasks and modify their behaviours in response to environmental cues and interactions with other colony members. Role-switching plasticity is essential for maintaining an equitable distribution of responsibilities, ensuring that critical activities such as brood care, foraging, and hive upkeep are consistently managed. Furthermore, it allows the colony to effectively adapt to disturbances, such as the loss of key members or fluctuations in resource availability, by redistributing tasks among the remaining workers. This flexibility is crucial for enhancing the colony's performance, resilience, and survival in an ever-changing environment.

Temporal polyethism: Temporal polyethism plasticity plays a crucial role in the task distribution among stingless bees, highlighting their capacity to modify their responsibilities following age-related transformations and the temporal requirements of the colony. This adaptability enables bees to progress through various tasks as they age, with younger members typically engaging in activities within the hive, such as caring for the brood and maintaining the nest, before gradually transitioning to foraging and other external duties as they grow older. This age-dependent progression is supported by the bees' ability to modify their behaviours and roles in response to both internal signals, such as physiological changes, and external influences, such as fluctuations in resource availability or the needs of the colony. Temporal polyethism ensures that bees are deployed most effectively throughout their lives, with younger bees concentrating on tasks that necessitate less foraging experience. In comparison, older bees assume more intricate roles that require more outstanding expertise and skill. This adaptable task allocation contributes to the overall efficiency and productivity of the colony, enabling it to optimise its workforce in alignment with the varying demands of different life stages and environmental conditions.

2. **Individual responsiveness:**

Task abandonment and adoption: The processes of task abandonment and adoption are essential elements of individual responsiveness in the flexible allocation of tasks among stingless bees. These behaviours underscore the bees' capacity to adjust their roles dynamically per immediate needs and fluctuating circumstances. Task abandonment occurs when a bee stops its current function due to changes in the colony's demands or external influences, such as a lack of resources or threats. This adaptability enables the bee to take on new tasks that better fit the colony's current requirements. On the other hand, task adoption refers to a bee assuming new duties that were either previously handled by others or are newly necessary. Various factors can affect this shift, including signals from fellow bees, environmental feedback, or the individual bee's experiences and skills. For example, if a sudden increase in nectar necessitates additional processing and storage, bees may leave their existing roles to engage in foraging or nectar-handling activities. Such responsive behaviour allows the colony to swiftly adapt to changes and maintain an adequate distribution of tasks, thereby improving overall resilience and efficiency. By effectively balancing the abandonment of tasks with the prompt adoption of new roles, stingless bees exhibit exceptional adaptability and coordination, vital for sustaining colony operations and responding to ever-changing environmental conditions.

Workforce reserve: The concept of workforce reserve plays a crucial role in enhancing the adaptability of stingless bees in task distribution. This term denotes the group of bees within a colony that are not actively engaged in essential duties but are ready to be deployed as necessary. These reserve workers can swiftly shift into various functions, including foraging, caring for the brood, or maintaining the hive in response to unexpected changes or heightened demands. A workforce reserve allows the colony to effectively handle variations in task requirements, particularly during increased brood rearing or when there is an abrupt decline in forager numbers. This reserve is sustained through task rotation, where bees periodically change roles to avoid monotony and fatigue and through the colony's capacity to recruit and reassign workers according to current needs. By utilising this reserve, stingless bees can quickly respond to internal and external challenges, ensuring that all critical tasks are carried out efficiently and that the colony remains robust and adaptable to changing circumstances.

CHAPTER 6

SWARMING OF INDIAN STINGLESS BEES

What is swarming?

Swarming is a natural reproductive process in which a colony of bees, including honey bees and stingless bees, divides to create new colonies. It occurs when a portion of the colony and the old queen leaves the original nest, searching for a new nesting site. Swarming typically happens during the reproductive season when colony populations are high and environmental conditions are favourable.

Events of swarming

During the swarming of stingless bees, scout workers search for a good nesting site near the mother nest. Once they select a suitable nesting site, workers clean and transport food reserves and hive material to the new nest. They begin to construct the new nest, and once completed, more workers, together with a virgin queen from the old nest, will migrate to the new nest. When the daughter colony became virtually independent of the mother colony (approximately one week after swarming), the proportion of workers placed in it was around 30%. The mass flight of males lasted for two days after the swarm. The virgin queen will make her nuptial flight shortly after landing at the new nest. In most species, mating between a new queen and drones usually occurs outside the nest (Inoue et al. 1984b). The contact between mother and daughter nest can endure from a few days to several months, depending on the availability of materials and the bee species (Inoue et al. 1984a; Van Veen and Sommeijer 2000). Different steps of swarming are described below:

1. **Preparation phase:**

Colony expansion: As the colony population increases and resources become abundant, the colony may outgrow its current nesting site.

Queen rearing: The colony may start rearing new queens by constructing special queen cells within the nest. These queen cells are larger than regular brood cells and are used to rear potential queen replacements.

2. **Swarm preparation:**

Reduction in queen's egg-laying: As new queens develop and mature, the egg-laying activity of the old queen typically decreases.

Resource accumulation: Worker bees start storing honey and pollen reserves in preparation for the swarming process.

3. **Scouting and selection:**

Scout workers: Scout bees search for potential new nesting sites near the original colony. They evaluate factors such as size, protection from predators, and suitability for brood rearing.

Decision-making: Once a suitable nesting site is identified, scout bees communicate their findings to the colony and reach a consensus on the new location.

4. **Swarm departure:**

Swarm formation: When conditions are optimal, usually on warm and sunny days, a portion of the colony, including the old queen and a group of worker bees, leaves the original nest in a swarm.

Swarm behaviour: The swarm flies together in a swirling mass, emitting buzzing sounds as they search for a suitable landing spot.

5. **Settling in a new nest:**

Nest establishment: Once a suitable site is found, the swarm settles and establishes a new nest. This could be inside tree hollows, building cavities, or other protected locations.

Nest construction: Worker bees begin constructing the new nest using wax, resin, and other plant materials.

6. **Colony establishment:**

Queen's role: The old queen takes up residence in the new nest and resumes egg-laying to establish the new colony.

Worker activities: Worker bees engage in nest construction, foraging for food, caring for brood, and other colony maintenance tasks to ensure the success of the new colony.

7. **Remaining colony:**

After swarming, the original colony, now without the old queen, continues its activities under the care of the newly emerged queens and remaining worker bees.

Period of swarming

Swarming generally occurs during honey flow (Chauhan and Singh 2020). In Indian states, it mainly occurs between November and March. When the amount of incoming pollen and nectar into a nest is high, the queen receives stimuli to brood rearing, and several queen cells are formed. Ultimately, virgin queens are developed, and the colony goes to swarming.

Cause of swarming

1. **Overcrowding and lack of ventilation**:

Due to colony growth and limited nest space, the colony became overcrowded, and they needed to swarm to balance colony size.

2. **Greater floral resource availability**:

When a large amount of pollen and nectar flow into the nest, the queen reaches a peak brood-rearing stage. The number of bees within the colony increases, and virgin queens also develop, sequentially promoting swarming.

3. **Presence of old queen**:

Bee communication mainly depends on the pheromones produced by the queens. When the queen is old, the workers receive fewer pheromones. This stimulates them to make new queen cells and queens, which ultimately lay down swarms.

Aggressive swarming or fighting swarms

The workers aggregate in front of the hives, and hundreds of bees die after aggregation. This behaviour is shown in different stingless bee species like *Trigona carbonaria* and *Tetragonula iridipennis*. In *Tetragonula iridipennis*, fighting often leaves the colonies weak or sometimes mass death of the robbed colony. The reasons behind the fighting of swarms are as follows:

1. Aggressive swarming occurs due to a nearby colony's invasion of a new swamp, which unfortunately selected this nest as its nesting site. Worker bees clutch together and bite each other, which finally leads to their fetal death.
2. Stealing the stored foodstuffs of another colony due to cleptobiosis may lead to fighting swarms (inter-colony attacks).

Problems of swarming in managed stingless bee colonies

1. **Loss of colony productivity**:

When a colony swarms, a portion of the workforce, including foragers and nurse bees, leaves with the old queen to establish a new colony. This significantly reduces the workforce within the original colony, leading to decreased productivity in honey production, pollen collection, and brood rearing.

2. **Reduction in genetic diversity**:

Swarming can lead to colony fragmentation and the establishment of new colonies with limited genetic diversity. Over time, this can result in inbreeding and reduced colony health and productivity.

3. **Decrease hive yield**:

Swarming often occurs during the peak honey-producing season. As a result, beekeepers may experience a reduction in honey yield from swarmed colonies, which can impact their economic viability.

4. **Queen loss**:

Swarming involves the departure of the old queen with a portion of the colony's workers. In some cases, the beekeeper may lose the original queen, which can delay the colony's recovery and require the introduction of a new queen or the promotion of a virgin queen.

5. **Colonizing undesirable locations**:

Swarms may settle in undesirable locations, such as inside buildings, hollow trees, or other inaccessible areas. Retrieving swarms from such locations can be challenging and may require specialised equipment and expertise.

6. **Disruption of management practices**:

Swarming can disrupt the beekeeper's management practices, requiring additional time and effort to reestablish colonies, redistribute resources, and monitor the health and productivity of the remaining colonies.

7. **Potential of aggressive behaviour**:

Swarming bees may exhibit defensive behaviour, particularly during swarming and while scouting for a new nesting site. This can pose a risk to beekeepers and bystanders near swarming colonies.

Control of swarming in managed stingless bee colonies

Controlling swarming in managed stingless bee colonies is essential for maintaining colony productivity and preventing the loss of bees. Here are some strategies that beekeepers can use to manage swarming:

1. **Providing adequate space**:

Swarming often occurs when colonies become overcrowded. To accommodate the growing colony population, regularly inspect the hives and add new hive boxes or frames.

2. **Monitor colony population**:

Monitor colony population dynamics and be alert for signs of overcrowding, such as congestion at hive entrances, reduced foraging activity, or the presence of queen cells.

3. **Artificially colony division**:

Proactively divide colonies by creating artificial splits or divisions. This involves separating a portion of the colony, including the queen and some worker bees, into a new hive box. This can help alleviate overcrowding and reduce the likelihood of swarming.

4. **Swarm traps**:

Install swarm traps near the apiary to capture swarms before they settle in undesirable locations. Swarm traps can be baited with pheromones or attractants to attract swarming bees.

5. **Queen suppression technique**:

Implement techniques to suppress queen production, such as removing queen cells or introducing queen excluders to prevent the queen from laying eggs in certain hive areas.

6. **Provide adequate nutrition**:

Ensure colonies have access to diverse floral resources to meet their nutritional needs. Adequate nutrition can help maintain colony health and reduce the likelihood of swarming.

7. **Monitor environmental conditions**:

Be aware of environmental factors that may trigger swarming, such as temperature fluctuations, rainfall patterns, or the availability of flowering plants. Manage colonies accordingly during periods of environmental stress.

8. **Regular hive inspection**:

Conduct regular hive inspections to assess colony health, identify potential swarming triggers, and take proactive measures to prevent swarming.

9. **Selective breeding**:

Select and breed stingless bee colonies with traits associated with reduced swarming propensity, such as low swarming tendency, high honey production, and good disease resistance.

CHAPTER 7

THREATS, COLONY DEFENCE, AND CONSERVATION OF INDIAN STINGLESS BEES

Threats for stingless bees

Stingless bees face multiple threats, including habitat loss, pesticide exposure, predation, diseases, climate change, pollution, competition with invasive species, and poor awareness about stingless bees. These factors can weaken stingless bee colonies and lead to population declines.

1. Habitat distraction

Habitat loss for stingless bees is primarily driven by human activities that alter or destroy natural landscapes. Some critical reasons behind habitat loss for bees are given below:

Lost nesting substrate:

Stingless bee colonies are perennial and do not leave their old nest without vigorous threats. Nest-containing old trees may perish due to ageing or being cut out by humans, damaging the stingless bee colonies. Corroding walls containing nests may naturally collapse after a certain period or be broken by humans, causing the nests to be damaged (Figure 7.1). Sometimes, the nest-containing walls were repaired, and the nests were damaged (Figure 7.2).

Deforestation:

Large-scale forest clearing for agriculture, urbanisation, and infrastructure development removes crucial bee habitats. Forests often provide diverse floral resources and nesting sites for many bee species, including stingless bees.

Urbanisation:

The expansion of cities and towns consumes natural habitats, including meadows, grasslands, and scrublands, which are important for wild bee populations. Urban areas typically offer less floral diversity and nesting opportunities than natural landscapes.

Agricultural expansion:

The conversion of natural habitats into agricultural land, such as large-scale monoculture farming, reduces floral diversity and disrupts the availability of suitable foraging areas for bees. Intensive agricultural practices often involve pesticides and herbicides, further harming bee populations.

Infrastructure development:

Construction of roads, highways, dams, and other infrastructure projects can fragment habitats and disrupt the natural corridors bees rely on for foraging and dispersal.

Land use practices:

Unsustainable land management practices such as overgrazing, soil erosion, and habitat fragmentation can degrade natural habitats and reduce the suitability of landscapes for bees.

Figure 7.1. (A–B) Showing wild nests of Indian stingless bees. (C) Breakdown of the wall causes the loss of the wild nests.

Figure 7.2. (A) A wild nest of stingless bees within an old wall. (B) The repaired wall does not show the nest. (C) An enlarged view of the repaired wall shows the sealed and damaged nest.

2. Monoculture and agrochemicals

Monoculture, cultivating a single crop species over large areas, poses several threats to stingless bees and other pollinators.

Reduced floral diversity:

Monoculture farming typically involves planting vast expanses of a single crop species, which limits the availability of diverse floral resources for bees. Stingless bees rely on various nectar and pollen sources for their nutrition, and monoculture landscapes often cannot provide sufficient food throughout the bees' foraging season.

Limited foraging opportunities:

Monoculture crops may only bloom briefly, providing stingless bees a limited window of foraging opportunities. Once the crop flowering period ends, bees may struggle to find alternative food sources, leading to nutritional stress and colony decline.

Pesticide exposure:

Monoculture farming relies heavily on pesticides to control pests and diseases that can damage single-crop species. These pesticides can inadvertently harm stingless bees through direct contact, ingesting contaminated nectar and pollen, or secondary poisoning from consuming contaminated prey or floral resources.

Loss of nesting sites:

Monoculture landscapes typically lack the diverse vegetation structures and nesting substrates that stingless bees require. These bees often nest in tree hollows, soil cavities, or other natural crevices, but monoculture fields offer limited nesting opportunities, leading to habitat loss and population decline.

Genetic homogeneity:

Monoculture farming can result in genetic homogeneity within crop populations, which may reduce the availability of certain floral traits that stingless bees prefer. This lack of floral diversity can further limit the nutritional quality of the nectar and pollen available to bees.

Increased vulnerability to diseases and parasites:

Monoculture cropping systems can create ideal conditions for the proliferation of bee pathogens and parasites. When bees are concentrated in large numbers around monoculture fields, diseases and parasites can spread more easily among colonies, leading to higher infection rates and colony collapse.

Agrochemicals, including pesticides and fertilisers, threaten stingless bees and pollinators. These can negatively impact stingless bee populations in the following ways:

Pesticide toxicity:

Many pesticides used in agriculture, such as insecticides, fungicides, and herbicides, are toxic to bees. Even at sublethal doses, these chemicals can impair the foraging behaviour, navigation abilities, and overall health of stingless bees. Pesticides can kill bees upon contact or ingestion, leading to immediate mortality.

Sublethal effects:

Even if not lethal, exposure to sublethal doses of pesticides can have detrimental effects on stingless bees. These effects may include impaired learning and memory, reduced foraging efficiency, decreased reproductive success, and compromised immune function, making bees more vulnerable to diseases and parasites.

Residue accumulation:

Agrochemical residues can accumulate in nectar, pollen, and other floral resources that stingless bees collect and consume. Continuous exposure to pesticide residues can lead to chronic poisoning and long-term health issues for bee colonies.

Impact on larvae and queens:

Pesticide exposure can affect all life stages of stingless bees, including larvae and queens. Larvae fed with contaminated provisions may experience developmental abnormalities or reduced survival rates, while queens exposed to pesticides may produce fewer or weaker offspring, leading to colony decline over time.

Loss of floral resources:

Herbicides that control weeds in agricultural fields can inadvertently eliminate flowering plants that serve as food sources for stingless bees. Reductions in floral diversity and abundance can lead to nutritional stress for bees and negatively impact colony health and productivity.

Synergistic effects:

The simultaneous exposure of stingless bees to multiple agrochemicals and other environmental stressors, such as habitat loss and climate change, can result in synergistic effects that exacerbate the harmful impacts on bee populations. These interactions may amplify the toxicity of individual chemicals and increase the overall risk to bee health.

3. Pests of stingless bees

Stingless bees face threats from various pests, including wax moths, small hive beetles, ants, mites, birds, and pathogens. These pests can weaken colonies by damaging combs, stealing food, or spreading disease. Pests pose significant threats to stingless bees, impacting their colonies in various ways:

Disruption of hive functioning:

Pests like wax moths, small hive beetles, and ants can disrupt the normal functioning of stingless bee hives by invading the colony, consuming resources, and causing chaos among the bees. This disruption can lead to decreased productivity and efficiency within the hive.

Weakening of bee population:

Pests often target the resources essential for the survival of stingless bees, such as pollen, honey, and developing brood. Continuous feeding by pests like wax moth larvae and small hive beetle larvae can weaken the bee population by depriving them of essential nutrients and resources needed for growth and reproduction.

Physical damage to comb and hive infrastructure:

Wax moths, small hive beetles, and other pests can cause physical damage to the hive infrastructure by burrowing into comb structures, tunnelling through honey stores, and creating pathways for other pests to invade. This damage weakens the hive's structural integrity and can collapse comb structures, further exacerbating the colony's vulnerability.

Predation and stress:

Predatory pests such as birds and mammals may directly attack stingless bee colonies, preying on bees, honey, and larvae. This predation leads to direct losses and induces stress within the colony, making them more susceptible to other threats, such as diseases and environmental stressors.

Spread of diseases and pathogens:

Some pests, like mites, can be vectors for diseases and pathogens that affect stingless bees. As these pests infest the colony and feed on bee hemolymph or bodily fluids, they can introduce and spread pathogens, leading to increased morbidity and mortality among the bee population.

Here, we briefly documented some pests for stingless bees, which are illustrated below:

Ants

Ants are common predators of stingless bees in tropical and subtropical regions. They attack the nests in mass and take out virtually everything, viz. honey, pollen, adult bees, and brood. The weak colonies of stingless bees are more susceptible to ant attacks. *Solenopsis geminata* Wheeler enters into the nests of weak colonies of *Tetragonula iridipennis* (Raakhee 2000) or at the initial stages when the colonies were just divided. Weaver ants (*Oecophylla smaragdina*) also attack adult stingless bees. Damaged or robbed nests are frequently attacked by ants. Ants can pose a threat to stingless bees in various ways:

- **Predation:** Ants may prey on stingless bee larvae, pupae, or adult bees. They can enter the hives for food and attack the brood or weaker bees.
- **Resource competition:** Ants compete with stingless bees for resources such as nectar, pollen, and nesting sites. This competition can stress the bees and impact their foraging efficiency and overall health.
- **Nest invasion:** Ants may invade stingless bee colonies, causing disturbance and potentially damaging the nest structure. They may steal stored food, disrupt the brood, or even kill the queen, leading to colony collapse.
- **Chemical warfare:** Some ant species produce chemicals or pheromones that deter or repel stingless bees. These chemicals can disrupt communication among bees or make the nest environment inhospitable.

- **Disruption of foraging:** Ants may create barriers or trails that interfere with the stingless bees' foraging activities, making it harder for them to collect resources efficiently.

Bees

Bees, particularly bigger species such as leafcutter bees (*Megachile* spp.) can occasionally prey on smaller stingless bees, especially when resources are low. Competition for foods (nectar and pollen) and the demand for nesting material (Figure 7.3) typically drive these interactions, with larger bees using their bulk and strength to dominate or even kill smaller bees to get access to food resources and nesting materials. Stronger solitary bees frequently attack deteriorating stingless bee colonies to collect foodstuff (honey, pollen, and larvae). Sometimes, honeybees also attack a newly settled stingless bee colony to collect honey.

Figure 7.3. A leafcutter bee (*Megachile disjuncta*) collecting propolis from a stingless bee colony.

Beetles

The Nitidulid beetles are important destructive pests of stingless bees, causing damage to brood cells, honey, and pollen pots. Small hive beetle larvae feed on honey, pollen, and brood within the hive, causing damage to comb structures and leading to the fermentation of stored honey. Infestations can cause hive abandonment and ultimately result in a collapse of the colony if left unmanaged. Many stingless bee species (e.g.,

Meliponula bocandei Cockerell, *Meliponula ferruginea* Cockerell, and *Hypotrigona gribodoi* Magretti) in Kenya are vulnerable to invasion by *Aethina grubs* (Kiatoko 2012) and also recorded other beetles (e.g. *Rhizoplatys mucronatus* Beauvois and *Tenebroidei mauritanicus* L.) from the food pots of stingless bee colonies. Hive beetles (e.g., *Carpophilus* spp.) and sap beetles (*Brachypeplus* spp.) were found to attack Australian stingless bees (Spooner-Hart et al. 2018). However, specific beetle attacks have yet to be reported for the Indian stingless bees.

Birds

Birds catch the adult stingless bees from the nest entrances or while foraging on flowers. The common house sparrow is identified as a minor predator of stingless bees in India. As Indian stingless bees are small-sized, large birds do not prefer them as profitable prey. The African blue flycatcher bird (*Elminia longicauda* Antinori) frequently hunts foragers of stingless bees (*Melipona ferruginea*) in Kenya (Kiatoko 2012).

Figure 7.4. (A) Stingless bee forager trapped within a spider web, and (B) a spider directly caught a stingless bee forager.

Bugs

Bugs pose multifaceted threats to stingless bees. They can directly invade hives, prey on bees, compete for floral resources, and act as disease vectors. The adults and nymphs of bugs are observed around the hives, crevices, and on the tree trunks in the vicinity of stingless bee colonies. The reduviid bug (*Acanthaspis siva* Distant) was reported as a predator of stingless bees in Kerala, India (Premila et al. 2013). Outside India, *Apiomerus pillipes* Santschi is a predator of Brazilian stingless bees (Silva and Gl-Santana 2004) and *Pahabengkakia piliceps* Miller is a predator of Thai stingless bees (Wattanachaiyingcharoen and Jongjitvimol 2007).

Flies

Different dipteran flies act as pests for several stingless bee species. For example, *Hermetia illucens* L. on Indian stingless bee species (Nisha 2002), *Ceriana ornata* Macquar on Australian stingless bees (Klumpp 2007), *Mealonctia sinistia* Borgmeier on Brazilian stingless bees (*Nannatrigona postica* Latreille) in Brazil (Simoes et al. 1980) and *Pseudohypocerrspo* sp. on American stingless bees (Sommeijer 1999). *Hermetia illucens* thrives on decaying organic matter in the nest of *Tetragonula iridipennis*. Both the maggots and adults attack stingless bee colonies. The larvae voraciously fed on the honey, pollen, and brood of stingless bees (Nisha 2002).

Lizards

The Indian stingless bees (*Tetragonula iridipennis*) are frequently attacked by lizards. Some small-sized lizards attack at the nest entrance of stingless bee colonies, while most large-sized lizards catch and eat adult bees in the field when they forage to collect floral rewards. Generally, lizards do not enter into stingless bee nests to catch adult bees.

Mites

Many mites are found in the nests of stingless bees. Mites were reported for Indian stingless bee nests (Vijaykumar et al. 2013) and also for Brazilian stingless bees (Flechtmann et al. 1974) and Malaysian stingless bees (Baker and Delfinado 1985). For *Tetragonula iridipennis* colonies, several mite species (e.g., *Neocypholaelaps malayensis*: Delfindo-Baker et al. 1989; *Hypoaspis* sp.: Smiley et al. 1996; *Carpoglyphus lactis* L.: Vijaykumar et al. 2013) were recorded. Certain mites parasitize stingless bees. They can attach themselves to adult bees or their brood (larvae and pupae) and feed on their fluids or tissues. This parasitic association can damage individual bees while jeopardizing the overall health of the colony.

Spiders

Different species of spiders were recorded as predators of stingless bees. Jayalekshmi (2015) documented eight spiders (e.g., *Crossopriza lyoni*, *Hersilia savignyi*, *Heteropoda venatora*, *Menemerus bivittatus*, *Neoscona nautica*, *Plexippus paykulli*, *Thomisus lobosus* and *Thomisus projectus*) as predators for *Tetragonula iridipennis* from Kerala, India. Most spiders build webs to catch prey, and stingless bees can inadvertently get caught in these webs while foraging for nectar or pollen (Figure 7.4). Once trapped, the bees may become easy prey for the spider. Sometimes spiders can make webs near the entrance of stingless bee nests. Some spiders don't build webs to catch prey; instead, they invade the nests of stingless bees and prey on adult bees and larvae.

4. Diseases of stingless bees

Stingless bees are not very susceptible to diseases, unlike honeybees. Only a few bacterial and fungal diseases are reported for stingless bees. Jayalekshmi (2015) reported a few fungal strains (e.g., *Aspergillus niger* and *Paecilomyces variotii*) from abandoned nests, food pots, and dead bees lying inside the hive floor. For *Tetragonula carbonaria*, Shanks brood disease was reported to be caused by a bacterium, *Lysinibacillus sphaericus* (Shanks et al. 2017).

5. Climate change

Climate change poses several significant threats to stingless bees and their habitats:

Altered flowering patterns:

Climate change can disrupt the timing of flowering in plants, leading to mismatches between the availability of floral resources and the foraging activities of stingless bees. Changes in temperature and precipitation patterns can affect the phenology of flowering plants, potentially reducing the abundance and diversity of nectar and pollen sources for bees.

Habitat loss and fragmentation:

Rising temperatures, changes in rainfall patterns, and extreme weather events associated with climate change can lead to habitat loss and fragmentation for stingless bees. Shifts in temperature and precipitation regimes may render certain habitats unsuitable for bees, forcing them to relocate or face local extinction. Habitat fragmentation can isolate bee populations, reducing gene flow and increasing the risk of inbreeding and genetic bottlenecks.

Increased frequency of extreme weather events:

Climate change is expected to increase the frequency and intensity of extreme weather events such as heatwaves, droughts, floods, and storms. These events can directly impact stingless bee colonies by causing mortality, damaging nesting sites, disrupting foraging activities, and reducing reproductive success.

Loss of floral diversity:

Climate change can alter plant species' distribution and abundance, leading to floral diversity shifts within ecosystems. Some plant species may decline or become locally extinct due to changing environmental conditions, reducing the availability of nectar and pollen resources for stingless bees and other pollinators.

Spread of invasive species and diseases:

Climate change can facilitate the range expansion of invasive species and the emergence of new pests and diseases that threaten stingless bees. Warmer temperatures and altered precipitation patterns may create more favourable conditions for invasive species to thrive, out-competing native flora and fauna and disrupting ecosystem dynamics.

Disruption of symbiotic relationships:

Stingless bees often rely on mutualistic relationships with specific plant species for food and nesting resources. Climate change-induced shifts in plant distributions and phenology can disrupt these symbiotic relationships, negatively impacting their survival and reproductive success.

Reduced resilience to other stressors:

Climate change can reduce the resilience of stingless bee populations to other stressors such as habitat loss, pesticide exposure, and diseases. Bees already facing multiple threats may be less able to cope with additional environmental challenges brought about by climate change, leading to population declines and local extinctions.

6. Pollution

Pollution from various sources, including industrial activities, vehicle emissions, and agricultural runoff, can degrade bee habitats and contaminate floral resources. Pollution reduces floral diversity and can harm bee health and reproductive success. Therefore, Pollution poses several significant threats to stingless bees, impacting their health, behaviour, and overall survival.

Contaminated floral resources:

Pollution from various sources, including industrial activities, vehicle emissions, and agricultural practices, can contaminate nectar and pollen from flowers. Stingless bees rely on these floral resources for their nutrition, and exposure to contaminated food sources can lead to poisoning and adverse health effects.

Direct toxicity:

Pollutants such as heavy metals, pesticides, and industrial chemicals can harm stingless bees upon contact, ingestion, or inhalation. These toxic substances can interfere with bee physiology, impairing vital functions such as respiration, digestion, and reproduction. Prolonged exposure to high levels of pollutants may lead to acute poisoning and mortality among bee populations.

Impaired foraging behaviour:

Pollution can disrupt the foraging behaviour of stingless bees by altering the chemical composition of floral scents and nectars. Bees use scent cues to locate and identify flowers, but pollution can mask or distort these cues, making it difficult for bees to find suitable foraging sites. As a result, bees may spend more time and energy searching for food, reducing foraging efficiency and colony productivity.

Weakened immune systems:

Exposure to pollutants can weaken the immune systems of stingless bees, making them more susceptible to diseases, parasites, and other environmental stressors. Pollutants may compromise bee immune function by disrupting cellular processes, inhibiting detoxification mechanisms, and promoting oxidative stress, leaving bees vulnerable to infections and pathogens.

Developmental abnormalities:

Pollution can interfere with the development of stingless bee larvae, leading to developmental abnormalities and reduced survival rates. Larvae exposed to high levels of pollutants may experience morphological deformities, growth retardation, and impaired physiological functions, ultimately affecting the overall health and fitness of bee colonies.

Behavioural disruption:

Pollution, such as noise and light pollution, can disrupt the natural behaviours of stingless bees. For example, loud noises from human activities can stress bees, interfere with their communication and navigation abilities, and disrupt their nesting behaviours. Similarly, artificial light at night can confuse bees and disrupt their circadian rhythms, impacting their foraging patterns and colony dynamics.

Population declines:

Cumulative exposure to pollution over time can lead to population declines and local extinctions of stingless bees. Pollution-related stressors and other threats, such as habitat loss, climate change, and invasive species, can push bee populations beyond their tolerance limits, resulting in significant declines in abundance and diversity.

7. Competition with invasive species

Invasive alien species pose several threats to stingless bees and their habitats:

Competition for resources:

Invasive alien species, such as other bee species or non-native plants, may compete with stingless bees for floral resources such as nectar and pollen. Increased competition can reduce food availability for stingless bees, leading to nutritional stress and decreased colony health and productivity.

Habitat modification:

Some invasive alien species can modify or degrade native habitats, altering the availability of suitable nesting sites and foraging areas for stingless bees. For example, invasive plants may outcompete native vegetation, reducing floral diversity and habitat quality for bees.

Predation and parasitism:

Invasive species like predatory insects or parasitic mites can directly prey upon or parasitise stingless bee colonies. Predation and parasitism can cause significant mortality among bee populations, particularly if invasive species have no natural predators or control mechanisms in their new environment.

Transmission of diseases and pathogens:

Invasive alien species may introduce new diseases, parasites, and pathogens to native stingless bee populations. These pathogens can spread rapidly through bee colonies, leading to increased rates of infection, reduced colony fitness, and population declines.

Hybridisation and genetic pollution:

Invasive alien species may hybridise with native stingless bee species, leading to genetic introgression and loss of genetic diversity within native populations. Hybridisation can weaken the genetic integrity of native bee populations and reduce their ability to adapt to environmental changes and stressors.

Disruption of mutualistic relationships:

Invasive species can disrupt mutualistic relationships between stingless bees and native plants, which rely on each other for pollination and reproduction. For example, if invasive species outcompete native plants, the availability of floral resources for stingless bees may decline, leading to cascading effects on ecosystem dynamics and biodiversity.

Altered ecosystem dynamics:

The introduction of invasive alien species can disrupt natural ecosystem dynamics, leading to shifts in species composition, community structure, and ecological interactions. These changes can indirectly impact stingless bees by altering the availability of resources, habitat suitability, and overall ecosystem resilience.

8. Lack of awareness

The lack of awareness about stingless bees poses significant threats to their survival. Without understanding their importance, people may inadvertently harm bee populations through habitat destruction, pesticide use, and mismanagement of landscapes. This can lead to habitat loss, population declines, and underestimation of the valuable ecosystem services stingless bees provide. To address these threats, educational outreach efforts are crucial to raise awareness among the public, policymakers, and land managers about the ecological significance of stingless bees and the actions needed to conserve them. Collaboration between various stakeholders is essential to implement effective conservation strategies and protect stingless bee populations and their habitats.

Defence mechanisms of stingless bees

Stingless bees employ a variety of defence mechanisms to protect their colonies. These include chemical defences through pheromones, physical defence, and the presence of guard bees at nest entrances. They also utilise their swarming behaviour and sacrificial actions as effective defensive mechanisms for their colonies. Colony size and strength may play a significant role in defence against predators and other threats. Overall, these mechanisms collectively ensure the survival of stingless bees in their natural habitats.

1. Chemical defence

Stingless bees produce and release various chemical compounds as defensive measures. These compounds can include alarm pheromones that signal danger to other colony members and chemical deterrents to ward off predators and intruders. Key aspects of their chemical defence mechanisms are given below:

Alarm pheromones:

Stingless bees release alarm pheromones when they perceive a threat to the colony. These pheromones serve as warning signals, alerting other colony members to the presence of danger. Alarm pheromones can trigger defensive responses and mobilise the colony to protect itself.

Chemical deterrents:

Stingless bees produce chemical compounds that act as deterrents against predators and intruders. These compounds may have repellent properties or cause irritation, discouraging potential threats from approaching the colony or its resources.

Communication:

Chemical communication is essential within stingless bee colonies for coordinating various activities, including defence. Through the release of specific chemicals, bees can convey information about resource availability, colony status, and potential threats, facilitating collective defence responses.

Nest protection:

Chemical compounds are often used to defend the nest structure. Stingless bees may coat the entrance or inner surfaces of their nests with substances that deter predators or inhibit the growth of harmful microorganisms, helping to maintain the integrity and security of the colony's home.

Colony recognition:

Chemical cues play a role in colony recognition, allowing stingless bees to distinguish between members of their own colony and outsiders. This recognition helps maintain cohesion within the colony and enables effective defence against intruders.

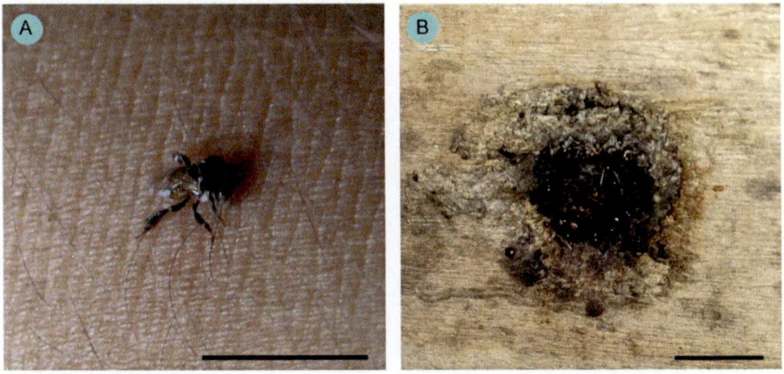

Figure 7.5. (A) A stingless bee biting on human skin, (B) an ant trapped into the propolis deposited at the nest entrance. Scale bars = 10 mm.

2. Physical defence

Stingless bees employ several physical defence strategies to protect their colonies from predators. While they lack a traditional stinger like honeybees, they have evolved other mechanisms to ward off threats. Here are some ways stingless bees physically defend against predators:

Biting:

Some species of stingless bees (e.g., *Tetragonula iridipennis*) use strong mandibles to deliver painful bites to predators (Figure 7.5A). Although not as powerful as a honeybee's sting, these bites can repel predators and help the overall defence effort.

Propolis deposition:

Stingless bees carry propolis with their mouth parts and deposit it on predators to drive them away. Sometimes, they also deposit propolis at their nest entrance to prevent predators from entering the nests (Figure 7.5B).

Blocking entrances:

Sometimes (especially during evening or other unfavourable conditions), stingless bees almost close their nest entrance and physically hinder predators or other intruders' access.

Mobbing behaviour:

When confronted with a threat, stingless bees may swarm and attack the predator. They cover the intruder with their bodies, forming a swarm capable of overwhelming and suffocating smaller predators or driving away larger ones due to their sheer numbers.

Heat generation:

Stingless bees can generate heat through muscle activity, which they may deploy defensively. When bees mob a predator, the temperature within the mass rises, making the intruder uncomfortable and potentially driving it away.

Intimidation displays:

Several stingless bee species adopt frightening actions when faced with a predator. They may buzz loudly, dart around the predator, or make quick movements to confuse or alarm the invader.

3. Guard bees mediated defence

Guard bees are specialised workers stationed at the entrance and guard the colonies. These guard bees monitor incoming and outgoing traffic, inspecting individuals who attempt to enter the colony. They are responsible for detecting and repelling any threats that may compromise the colony's safety. They provide colony defence through the following activities:

Detection:

Guard bees are stationed at the entrance of the nest or hive, where they continuously monitor incoming and outgoing traffic. They use their sensory organs to detect any unusual or potentially threatening stimuli, such as the presence of predators or intruders.

Communication:

Once a potential threat is detected, guard bees communicate this information to other colony members through various means, including chemical signals known as pheromones. These chemical signals help to alert other bees and mobilise them to respond to the threat collectively.

Physical defence:

Guard bees directly confront intruders or threats by physically blocking the entrance to the nest or hive and engaging in defensive behaviours. They may form a barrier at the entrance, preventing access to unauthorised individuals or organisms.

Alarm signals:

Guard bees emit alarm pheromones when they encounter a threat, signalling other colony members to prepare for defence. These alarm signals can rapidly spread throughout the colony, effectively triggering a coordinated response to repel the threat.

Harassing intruders:

Guard bees may actively harass intruders by biting, stinging (although stingless bees have reduced or no stinging capacity), or buzzing around them aggressively. This harassment can deter predators or intruders from approaching the colony further.

Patrolling:

Guard bees may also patrol the nest or hive perimeter, scanning the surrounding area for potential threats and intercepting them before reaching the colony.

4. Nest architecture-mediated defence

Stingless bees construct elaborate nests with complex entrances designed to deter intruders. Some species build their nests in hard-to-reach locations, providing additional protection against predators. Therefore, the nest architecture of stingless bees also contributes additional defence for their colonies. Nest architecture-mediated defence is mainly taken up by the following characteristics of nests:

Entrance design:

Stingless bee nests typically have a single narrow entrance or multiple small entrances. Specialised bees often guard these entrances, making controlling access and defending against intruders easier. The narrow entrance size may also restrict the entry of larger predators, providing an additional layer of defence.

Propolis-layering:

Stingless bees construct their nests using a mixture of plant resins, wax, and other materials known as propolis. This propolis is a protective barrier, sealing gaps and cracks in the nest walls to prevent predators, parasites, and pathogens from entering. The antimicrobial properties of propolis also help maintain the nest's hygiene and reduce the risk of disease.

Internal chambers:

Stingless bee nests are often divided into multiple internal chambers or compartments. These chambers serve various functions, including brood rearing, food storage, and nesting sites for the queen. The partitioning of the nest into compartments may help contain threats or localised disturbances, limiting their impact on the entire colony.

Tunneling and labyrinths:

Some stingless bees construct elaborate tunnel systems or labyrinth-like structures within their nests. These intricate layouts create barriers that can impede the progress of intruders and predators, making it more challenging for them to reach the brood or resources stored within the nest.

Thermal regulation:

The architecture of stingless bee nests also regulates colony temperature and humidity levels. Maintaining optimal environmental conditions is essential for the colony's health and survival, indirectly contributing to its defence by promoting the vigour and resilience of its members.

5. Swarming behaviour mediated defence

Swarming behaviour in stingless bees can indirectly contribute to their defence mechanisms in several ways. Swarming allows stingless bee colonies to expand their population and establish new nests. Larger colonies can better defend themselves against predators and other threats due to increased numbers of guard bees and worker bees available for defence. Swarming facilitates the dispersal of reproductive individuals (queens and drones) to new locations, promoting genetic diversity within stingless bee populations. Genetic diversity can enhance the colony's resilience to environmental stressors and disease, indirectly contributing to its defence against pathogens and other threats. Swarming enables stingless bee colonies to exploit new foraging areas and resources. By distributing foraging efforts across multiple locations, swarms can reduce competition for resources and minimise the risk of resource depletion, ensuring the colony's long-term survival and resilience. Swarming disperses the colony's members over a larger area, making it more challenging for predators to target the entire colony. Predators may find it difficult to locate and attack dispersed individuals, reducing the colony's overall predation risk. Swarms may relocate to new nesting sites in response to environmental threats or disturbances. This ability to relocate can help stingless bee colonies escape unfavourable conditions or avoid predation pressure, improving their survival prospects.

6. Sacrificial behaviour mediated defence

Sacrificial behaviour in stingless bees involves certain individuals or groups within the colony taking risks or making sacrifices to protect the overall welfare of the colony. This behaviour can contribute to the defence mechanisms of stingless bee colonies in several ways. When guard bees detect a threat, they emit alarm pheromones to alert other colony members. In doing so, these guard bees may attract the attention of predators or draw aggression toward themselves, acting as sacrificial individuals to protect the rest of the colony. Some stingless bees may engage in sacrificial distraction to divert predators' attention away from the colony. Sacrificial individuals may fly away from the nest, attracting the predator's focus and leading it away from the nest, allowing other colony members to escape or prepare for defence.

During times of scarcity or food competition, certain individuals within the colony may sacrifice their access to resources to ensure the survival of other colony members, particularly the brood or queen. This sacrificial behaviour helps maintain the colony's overall health and productivity, indirectly contributing to its defence.

7. Colony size and strength mediated defence

Colony size and strength are critical factors that mediate defence mechanisms in stingless bee colonies. Larger colonies have more worker bees available for defence. This means there are more bees to act as guards, patrol the colony perimeter, and respond to threats effectively. Stingless bee colonies can mount stronger and more coordinated defence efforts with a larger workforce. Larger colonies can maintain a higher number of guard bees at nest entrances. These guard bees play a crucial role in detecting and repelling threats. Larger colonies are often more resilient to threats due to their ability to produce swarms. Swarming behaviour allows colonies to expand and establish new nests, reducing the risk of predation or resource depletion in the original nest. By maintaining a large population size, stingless bee colonies can overwhelm predators with sheer numbers, reducing the likelihood of individual bees being targeted and increasing the chances of survival for the colony. Larger colonies have more bees available to defend these resources, making them more successful in controlling essential foraging areas and nesting sites.

Conservation strategies for stingless bees

Conservation of stingless bees is crucial for maintaining biodiversity, ecosystem stability, and agricultural productivity. Here are some strategies for conserving stingless bees:

1. Habitat preservation

Habitat preservation is a fundamental strategy for conserving stingless bees. It ensures the availability of suitable nesting sites, forage resources, and environmental conditions essential for their survival. Different modes of habitat preservation can be specifically tailored for stingless bee conservation.

Protecting natural habitats: Conserving natural habitats where stingless bees reside is paramount. This includes preserving diverse ecosystems such as tropical forests, grasslands, mangroves, and woodlands. These habitats provide nesting sites like tree hollows, fallen logs, and dense vegetation where stingless bees establish colonies.

Maintaining biodiversity: Biodiverse habitats support a variety of flowering plant species that provide nectar and pollen for stingless bees. Preserving plant diversity ensures a continuous and diverse supply of floral resources throughout the year, which is essential for stingless bee colonies' nutrition and survival.

Restoration efforts: Rehabilitating degraded habitats through reforestation, afforestation, and restoration of native vegetation can create suitable habitats for stingless bees. Restoring degraded areas helps recreate nesting sites and enhances the availability of forage resources, improving the overall habitat quality for stingless bees.

Protecting nesting sites: It is crucial to identify and conserve natural nesting sites for stingless bees. This includes preserving old trees with cavities, fallen logs, rock crevices, and other suitable nesting substrates. Protecting these nesting sites from disturbance, logging, and urbanisation helps maintain viable populations of stingless bees.

Buffer zones and corridors: Establishing buffer zones around protected areas and creating ecological corridors between fragmented habitats can enhance habitat connectivity for stingless bees. These buffer zones and corridors facilitate the movement of bees between different habitat patches, reducing isolation and genetic isolation risks.

Community-based conservation: It is essential to engage local communities in habitat preservation efforts. Collaborating with indigenous peoples, landowners, farmers, and stakeholders ensures the sustainable management of natural resources and promotes stewardship of habitats important for stingless bee conservation.

Incentives for conservation: Incentivizing landowners and communities to conserve natural habitats can encourage participation in conservation efforts. This may include financial incentives, payment for ecosystem services schemes, and recognition of traditional land management practices that support stingless bee habitats.

2. Promotion of native plants

Encouraging the planting of native flowering plants can provide essential forage resources for stingless bees. Creating pollinator-friendly gardens, green spaces, and agroforestry systems can help support healthy populations of stingless bees by ensuring a diverse and abundant supply of nectar and pollen. Here's how the promotion of native plants can contribute to stingless bee conservation:

Plant selection: We must identify and promote native plant species that are attractive to stingless bees. These plants should offer abundant nectar and pollen resources and be adapted to local environmental conditions. Consider factors such as flowering period, floral morphology, and nutritional value to ensure a diverse and continuous supply of resources throughout the year.

Pollinator gardens: Encouraging the establishment of native flowering plants that support stingless bees remains fruitful in raising the abundance of bees. These gardens can be created in various settings, including urban areas, parks, schools, and community spaces, providing additional forage resources and habitat for stingless bees.

Agroforestry systems: Promoting agroforestry practices that incorporate native trees and shrubs can increase the number of stingless bees. Agroforestry systems, such as shade-grown coffee, cacao, and fruit orchards, can enhance biodiversity, improve soil health, and provide valuable habitat for stingless bees while supporting sustainable agricultural production.

Landscaping guidelines: Developing guidelines for using native plants in public and private landscaping projects can help in bee management. It is necessary to encourage homeowners, businesses, and municipalities to choose native plant species for landscaping purposes, creating pollinator-friendly environments that benefit stingless bees and other native pollinators.

Educational campaigns: Raising awareness about the importance of native plants for bee conservation through educational campaigns, workshops, and outreach activities will help sustain stingless bees. The public needs to be informed about the role of native plants in providing food and habitat for stingless bees and the benefits of supporting native plant biodiversity.

Seed banks and nurseries: Establishing seed banks and native plant nurseries to propagate and distribute native plant species suitable for stingless bee habitats can enhance bee colonies. Work with local communities, botanical gardens, and conservation organisations to collect, propagate, and distribute native plant seeds and seedlings for restoration and landscaping projects.

Incentives and subsidies: By providing incentives and subsidies to landowners, farmers, and land managers for planting native vegetation that can support stingless bees. Financial incentives, tax breaks, and grants can encourage adopting practices that enhance habitat quality and promote native plant biodiversity.

3. Reducing pesticide use

Pesticides can harm stingless bees, including direct mortality and sublethal effects such as impaired foraging behaviour and reduced reproductive success. Therefore, reducing pesticide use is crucial for conserving stingless bees. Some strategies to reduce pesticide use and support stingless bee conservation are given below:

Integrated pest management (IPM): Implement IPM strategies that prioritise non-chemical pest control methods, such as crop rotation, habitat manipulation, biological control, and using pest-resistant varieties. This reduces reliance on pesticides and minimises their impact on pollinators like stingless bees.

Selective pesticide use: When pesticides are necessary, we need to choose the least harmful products to bees and other non-target organisms. Selective pesticides that have a minimal impact on beneficial insects, such as those with a short residual effect or those applied when bees are not actively foraging, can help protect stingless bees.

Alternative pest control methods: Exploring alternative pest control methods, such as pheromone traps, biopesticides, and mechanical methods like handpicking pests can eliminate negative impacts on beneficial insects. These methods can effectively control pests while minimising harm to stingless bees and other beneficial insects.

Separation between foraging time and pesticide use time: Most bees forage on flowers during the daytime, mainly from 7 a.m. to 5 p.m. The peak foraging time for Indian stingless bees is between 8 a.m. and 12 noon. If farmers use pesticides after 5 p.m., then the negative impacts of pesticides on stingless bees will be minimised.

Public awareness: Farmers, gardeners, and the general public must be educated about the importance of stingless bees and the potential harm of pesticides. It is also necessary to make stakeholders aware of alternative pest control methods and the benefits of reducing pesticide use for agricultural productivity and ecosystem health.

Policy support: We must advocate for policies promoting sustainable agricultural practices, reducing pesticide use, and protecting pollinators. It needs to encourage the development and implementation of regulations that restrict the use of harmful pesticides and incentivise the adoption of bee-friendly farming practices.

4. Artificial nest provision

Artificial nest provision involves creating or installing man-made structures designed to mimic natural nesting sites for stingless bees. Artificial nest provision can be an effective conservation strategy for stingless bees, especially in areas where natural nesting sites are limited or threatened. For an adequate artificial nest provision regarding stingless bee conservation, the following steps are needed to maintain:

Designing suitable nest structures: Artificial nests should replicate the features of natural nesting sites preferred by stingless bees. This may include hollow logs, wooden boxes with narrow entrance holes, or clay pots filled with suitable nesting materials such as resin, mud, or leaf litter.

Placement and installation: Artificial nests should be strategically placed where stingless bees will likely forage and establish colonies. This may include gardens, parks, agricultural landscapes, and natural reserves. Nest placement should consider factors such as sunlight exposure, protection from predators, and accessibility for beekeeping activities or research.

Monitoring and maintenance: Regularly monitoring artificial nests is essential to assess nesting activity, colony health, and overall success. This may involve periodic inspections to observe bee behaviour, check for signs of nesting activity, and determine nest conditions. Maintenance tasks such as cleaning, repairing damaged nests, and replacing nesting materials may also be necessary to ensure optimal nesting conditions.

Nest enhancement: Artificial nests can be enhanced to improve their attractiveness and suitability for stingless bees. This may include adding scent cues or pheromones to attract bees, providing nesting materials such as resin or mud, or incorporating features to regulate temperature and humidity within the nest structure.

Community engagement: Involving local communities in artificial nest provision initiatives can enhance conservation efforts and raise awareness about the importance of stingless bees. Community members can participate in nest installation, monitoring, and maintenance activities, fostering a sense of stewardship and ownership over bee conservation.

Scaling up conservation efforts: Artificial nest provision can be scaled to larger landscapes or regions to benefit multiple stingless bee species and populations. Collaborative efforts involving government agencies, conservation organisations, beekeepers, and landowners can facilitate the widespread deployment of artificial nests and promote bee-friendly land management practices.

Evaluation and adaptation: Continuous evaluation of artificial nest provision projects is essential to assess their effectiveness and identify opportunities for improvement. Monitoring key indicators such as nest occupancy rates, colony survival, and pollination services can help refine conservation strategies and adapt management practices.

5. Education and outreach

Raising awareness about the importance of stingless bees and their conservation needs is essential. Educational programs, workshops, and outreach activities can engage communities, farmers, policymakers, and the general public in efforts to conserve stingless bees and their habitats. Education and outreach play crucial roles in stingless bee conservation by raising awareness, fostering appreciation, and promoting actions to protect these essential pollinators. Several activities (given below) can increase awareness about stingless bee conservation.

Public awareness campaigns: Develop educational campaigns targeting the general public to raise awareness about the importance of stingless bees in ecosystems and agriculture. Use various media channels such as social media, television, radio, and print materials to disseminate information about stingless bees, their ecological roles, and conservation needs.

Educational institute programs: Integrate stingless bee conservation into school/college/university curricula at various levels to educate students about pollination ecology, biodiversity conservation, and sustainable agriculture. The educational institutes need to organise workshops, field trips, and hands-on activities to teach students about stingless bees and their habitats.

Community workshops: We need to organise community workshops, seminars, and events on stingless bee conservation. The experts will give talks, demonstrations, and practical training on beekeeping, nest installation, and habitat enhancement. We should also encourage community participation in conservation activities such as planting bee-friendly gardens and monitoring bee populations.

Demonstration sites: Establishing demonstration sites or bee reserves where people can observe stingless bees in their natural habitats and learn about conservation practices will promote stingless bee conservation. Additionally, providing interpretive signage, guided tours, and educational materials to inform visitors about stingless bee biology, behaviour, and conservation challenges will be effective for bee management.

Beekeeping courses: Several organisations offer training courses and workshops on stingless techniques for beekeepers, farmers, and enthusiasts. The courses cover hive management, colony propagation, honey harvesting, and pest management strategies that prioritise bee health and environmental sustainability.

Collaborative projects: Collaborate with local communities, conservation organisations, government agencies, and research institutions to develop and implement stingless bee conservation projects. Foster partnerships that leverage collective expertise, resources, and networks to address conservation challenges and achieve shared conservation goals.

Long-term engagement: Foster long-term engagement and commitment to stingless bee conservation by establishing networks, forums, and platforms for ongoing communication, collaboration, and knowledge exchange among stakeholders. Encourage individuals and organisations to take proactive steps to protect stingless bees and their habitats.

6. Research and monitoring

Research and monitoring are essential to stingless bee conservation efforts, providing valuable insights into bee ecology, population dynamics, threats, and conservation needs. Research and monitoring contribute to stingless bee conservation in several ways:

Understanding bee biology: Research helps scientists understand stingless bees' biology, behaviour, and ecological requirements. Studies may investigate nesting habits, foraging behaviour, reproductive biology, and interactions with other species. This knowledge forms the basis for effective conservation strategies tailored to the specific needs of stingless bee populations.

Assessing population status: Monitoring programs track changes in stingless bee populations over time, providing data on population trends, distribution, and abundance. Population assessments help identify areas of concern, prioritise conservation efforts, and evaluate the effectiveness of conservation interventions.

Identifying threats and risks: Research helps identify and quantify threats to stingless bee populations, such as habitat loss, fragmentation, pesticide exposure, disease, and climate change. By understanding the underlying causes of population declines, conservationists can develop targeted strategies to mitigate these threats and protect bee populations.

Habitat assessment and mapping: Research contributes to habitat assessment and mapping efforts to identify key habitats for stingless bees, including nesting sites, foraging areas, and floral resources. Mapping habitat distribution helps prioritise conservation areas, inform land use planning decisions, and guide habitat restoration initiatives.

Conservation genetics: Genetic research informs conservation strategies to preserve and enhance the genetic diversity of stingless bee populations. Conservation genetic approaches may include captive breeding programs, translocation efforts, genetic rescue, and habitat connectivity conservation to maintain gene flow among populations.

Pesticide impacts: Research investigates the effects of pesticides on stingless bees, including acute toxicity, sublethal effects, and long-term population consequences. Studies may assess pesticide exposure levels in bee habitats, evaluate pesticide residues in bee products, and identify pesticide formulations that pose the greatest risk to bee health.

Climate change adaptation: Research explores the impacts of climate change on stingless bee populations and their habitats. Studies may investigate how temperature, precipitation, and habitat suitability changes affect bee behaviour, phenology, and distribution. Understanding these impacts helps identify adaptation strategies to buffer bee populations against climate change effects.

7. Legislation and policy

Legislation and policy play crucial roles in stingless bee conservation by providing legal frameworks, regulatory measures, and incentives to protect bee populations and their habitats. The implementation of legislation to designate and protect critical habitats for stingless bees, including nesting sites, foraging areas, and floral resources. Protected area networks, conservation easements, and habitat corridors can safeguard key habitats from development, habitat destruction, and land use conversion. We need to enact laws and policies to conserve biodiversity, recognising the importance of stingless bees as pollinators and keystone species in ecosystems. Incorporating stingless bee conservation priorities into national biodiversity strategies, action plans, and conservation programs will strengthen their conservation. The regulation of the use of pesticides to minimise their impacts on stingless bees and other pollinators also remains effective in bee conservation. Further, we must develop pesticide registration processes that prioritise bee safety and require rigorous testing for pesticide toxicity to bees. Implementing pesticide application restrictions, buffer zones, and integrated pest management (IPM) practices to reduce pesticide exposure to bees is also used in stingless bee conservation. The integration of stingless bee conservation into land use planning processes at local, regional, and national levels is needed to consider and incorporate bee-friendly land management practices into zoning regulations, land development policies, and agricultural planning to maintain and enhance bee habitats. Governments develop policies and management plans to control invasive species that compete with or prey upon stingless bees. Governments also fund research, monitoring, and conservation projects focused on stingless bees. This will support scientific research to improve understanding bee ecology, behaviour, threats, and conservation needs. Collaborating with neighbouring countries and international organisations takes place to address transboundary issues affecting stingless bee conservation.

8. Community engagement and participation

Community engagement and participation are essential for the success of stingless bee conservation efforts. Conservationists can build support, foster stewardship, and implement effective conservation actions by involving local communities, indigenous groups, beekeepers, farmers, and other stakeholders. Community engagement and participation can contribute to stingless bee conservation in several ways:

Education and awareness: The conduct of educational outreach programs to raise awareness about the importance of stingless bees, their role in pollination, and the threats they face can diminish the extinction pressure.

Training: Organizing training on bee management can provide opportunities for community members, beekeepers, and farmers to learn about stingless bee identification, beekeeping techniques, habitat restoration, and sustainable land management practices. This will empower community members with the knowledge and skills to participate actively in bee conservation activities.

Traditional knowledge integration: We need to respect and incorporate indigenous knowledge and traditional practices related to stingless bees into conservation efforts. Collaborating with indigenous communities to document conventional beekeeping methods, identify important bee habitats, and integrate traditional ecological knowledge for conservation planning and decision-making.

Habitat restoration: Mobilising community members to participate in habitat restoration remains beneficial to stingless bees. We must organise tree planting events, community gardens, and restoration activities to create bee-friendly landscapes, restore degraded habitats, and provide forage resources for bees.

Nest installation and monitoring: Community members who engage in installing artificial nests and monitoring stingless bee populations can effectively enhance bee populations. For that, we need to train volunteers to identify suitable nesting sites, install structures, and monitor nesting activity. We also need to establish citizen science programs to involve community members in data collection, observation, and reporting on bee populations.

Pollinator gardens and green spaces: Encouraging community members to create pollinator-friendly gardens, green spaces, and urban habitats can support stingless bees and other pollinators. Community engagement can guide the selection of native plants, the creation of habitat features, and the avoidance of pesticide use to promote bee-friendly landscapes.

Cultural engagement: We must harness the power of culture, art, and storytelling to connect communities with stingless bee conservation. Considering this, it is demanding to organise cultural events, art exhibitions, and storytelling sessions that highlight the cultural significance of bees and inspire community members to take action to protect them.

Collaborative partnerships: Fostering collaborative partnerships between community organisations, conservation NGOs, government agencies, academic institutions, and other stakeholders can significantly increase stingless bee populations. By working together, diverse stakeholders can leverage their collective expertise and networks to implement comprehensive and effective stingless bee conservation initiatives.

CHAPTER 8

FORAGING ECOLOGY OF INDIAN STINGLESS BEES

Foragers

Foragers, in the context of stingless bees and, more broadly, in social insects, are specialised individuals within the colony who venture outside the nest to collect resources necessary for the colony's survival and growth.

Role of foragers

1. Nesting material collection:
2. Floral resources collection:
3. Sugary liquid collection:
4. Water: Necessary for various hive functions, including cooling the hive and diluting honey.

Characteristics of foragers

1. **Age and specialisation:** In many bee species, foraging is typically carried out by older worker bees. Younger workers usually perform tasks inside the hive before transitioning to foraging roles.
2. **Navigation skills:** Foragers have well-developed navigation skills, allowing them to accurately locate and return to the nest. They use visual landmarks, the sun's position, and possibly the Earth's magnetic field.
3. **Communication:** Foragers communicate the location of food sources to other colony members using chemical signals (pheromones) and specific behaviours. Unlike honeybees, stingless bees do not perform the "waggle dance" but have their own methods of conveying information.

Foraging activity

Stingless bees have diurnal foraging activity. Sometimes, young gourd bees can collect food resources after the sun is shed at dark if the source is very close to the nest entrance. Nectar foraging activity remains significant throughout the daytime. The number of incoming nectar foragers was low in the early morning (Figure 8.1). In colder seasons, fewer returning nectar foragers are in the late afternoon. In some time (during May), there are two peaks of returning nectar foragers. The number of incoming pollen foragers remains higher at 8.00–10.00 h. Then, pollen foragers gradually decrease with time (Figure 8.2). The least number of pollen foragers can be recorded in the late afternoon. Generally, most flowers open in the morning, so greater pollen availability remains for the foragers. In the late afternoon, pollen availability remains very low. So, pollen foraging activity also reduces at this time. Most of the time, the number of incoming latex and resin foraging is higher at 10.00–12.00 h. However, on warmer days (during March–April), peak activity is at 8.00–10.00 h, and on colder days (during January), it is at 12.00–16.00 h (Figure 8.3).

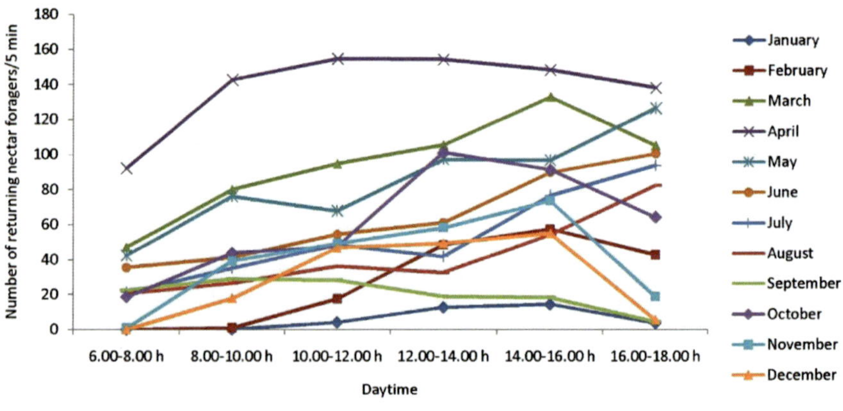

Figure 8.1. Daytime wise number of returning nectar foragers.

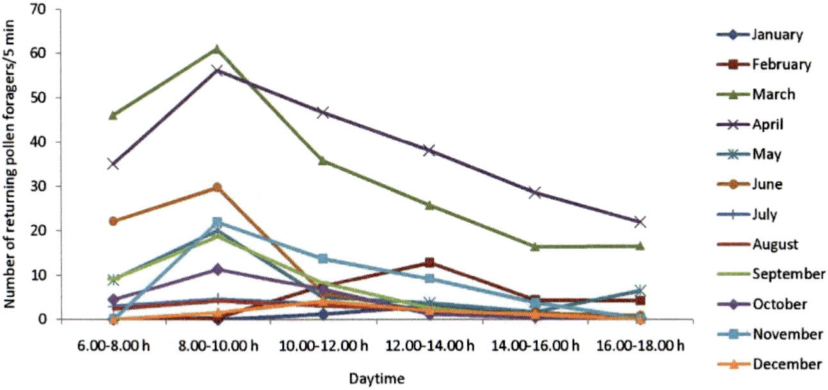

Figure 8.2. Daytime-wise number of returning pollen foragers.

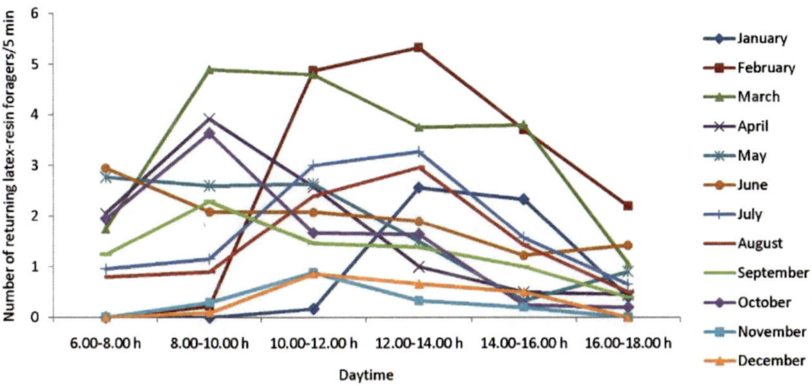

Figure 8.3. Daytime wise number of returning nectar foragers.

The foraging activity also varied month- and season-wise. The number of returning nectar foragers remains higher during spring–summer (i.e., March–May) and lower in the colder months of January and February (Figure 8.4). The pollen foragers were significantly higher during March–April than in the remaining months (Figure 8.5). This is due to favourable weather conditions to forage on flowers and greater availability of flower pollens. The number of latex and resin foragers is lower during colder months like November, December, and January than the rest of the month (Figure 8.6). In cold weather, their foraging activity remains less, and they actively forage for a short time.

Their foraging activity also varied according to Plant species, mainly influenced by the availability of floral resources. Foraging activity (considering abundance, foraging rate, and flower handling time) remains higher on some plants having to attract small-sized flowers (e.g., fennel, coriander) than the plants having medium to large-sized flowers (e.g., *Luffa aegyptiaca*). This can be determined by resource quality, quantity, and floral architecture. Their activity remains higher at 8.00–12.00 h on most plant species (Layek and Karmakar 2018).

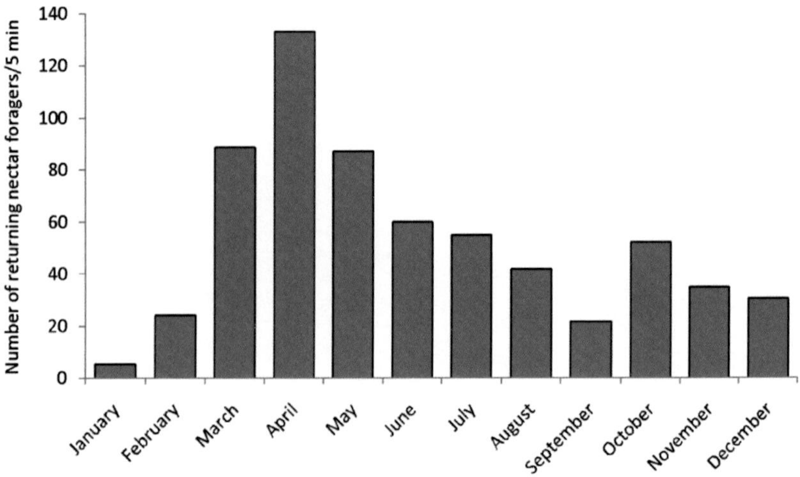

Figure 8.4. Daytime-wise number of returning nectar foragers.

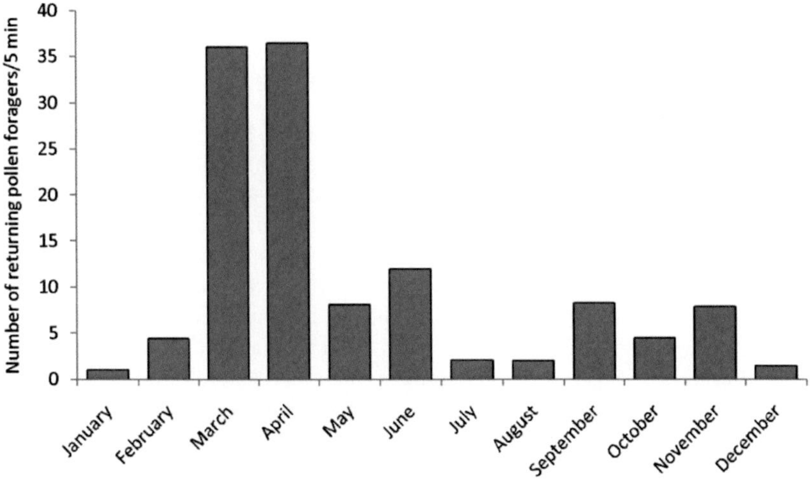

Figure 8.5. Daytime-wise number of returning pollen foragers.

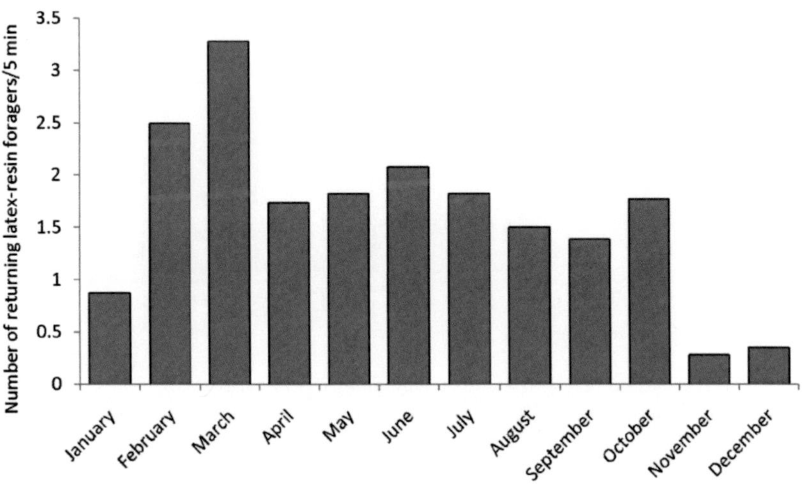

Figure 8.6. The number of returning latex and resin foragers is given month-wise.

Factors influencing foraging activity

1. Weather condition

Temperature: The number of incoming nectar foragers significantly correlates with temperature in both hot summer days (correlation coefficient, r = 0.68) and cold winter days (r = 0.54) (Figure 8.7). However, the number of returning pollen foragers and latex-resin foragers correlate less with temperature. Even at very high temperatures in summer, they show a negative correlation with temperature (pollen forager: r = -0.34; latex-resin forager: r = -0.16). During hot summer, their foraging activity remains higher at medium temperatures during the morning and late afternoon. At very high temperatures, their activity slows down. Floral nectar may dry out at very high temperatures (10.00–14.00 h) and be less available to the foragers. During the cold winter (December–January), foraging activity does not start early when temperatures remain low (12.46 ± 1.37). With rising temperatures, their activity also increases. Low temperature impairs their flight activity.

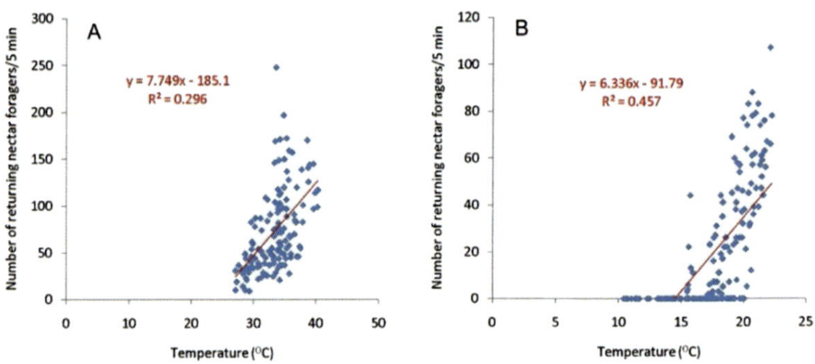

Figure 8.7. Scatter plots showing the relationship between temperature and the returning nectar foragers. (A) During hot summer, and (B) during cold winter.

Relative humidity: The number of returning nectar foragers correlates significantly negatively with humidity (r = -0.55) (Figure 8.8). The relative humidity remains high in the early morning, late afternoon, and rainy periods. During this time, nectar foraging activity remains low. During monsoons at high humidity, the nectar foraging activity remains low. Pollen-collecting activity and latex-resin foraging have a low correlation with humidity.

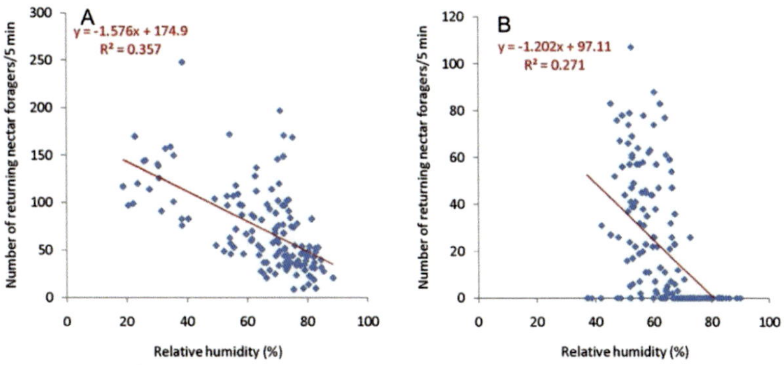

Figure 8.8. Scatter plots showing the relationship between humidity and the returning nectar foragers. (A) during hot summer, (B) during cold winter.

Rainfall: Bees' foraging behaviour is significantly impacted by rainfall, influencing their capacity to gather necessary resources like nectar and pollen. During rainy conditions, bees are often reluctant to leave their hives due to wet weather challenges. The rain can hinder their flight, as their wings become heavy and less aerodynamic when waterlogged, increasing the risk of injury or drowning. Additionally, raindrops can wash away nectar from flowers, reducing the availability of this vital food source. The drop in temperature during rainy weather also plays a role, as bees need a specific body temperature to sustain their flight and foraging activities. Colder temperatures can impede their ability to maintain the energy levels required for foraging, reducing activity. In addition, bees are highly responsive to the surrounding environment, and extended periods of rainfall can cause an increase in humidity inside the hive, potentially creating conditions conducive to mould and diseases, further straining the colony. This decrease in foraging affects the bees' immediate food consumption and has lasting implications for the hive's health and productivity. A decrease in foraging results in less nectar and pollen being returned to the hive, consequently impacting honey production and raising young bees. In both agricultural and natural settings, the reduced foraging activity of bees during rainy periods can lead to lower pollination rates, which in turn affects plant reproduction and the yield of crops reliant on bee pollination. The continuous or intense rain can cause a chain reaction, impacting the bees, the broader surroundings, and the agricultural methods that depend on their pollination services.

Light intensity: Nectar foraging activity correlates moderately with light intensity (Figure 8.9). In most daytime, nectar foraging activity positively correlated with light intensity. However, high light intensity in mid-day summer negatively influences nectar foraging activity. The pollen foragers and latex-resin foragers correlate less with light intensity in the daytime. During the early morning and late afternoon, when light intensity is very low, foraging activity is also very low. Light highly influences their navigation. Prey-predator interactions also depend on light. Flowering phenology, including nectar secretion and anther maturation, is also linked with light intensity. So, they avoid nocturnal foraging and highly early morning and evening.

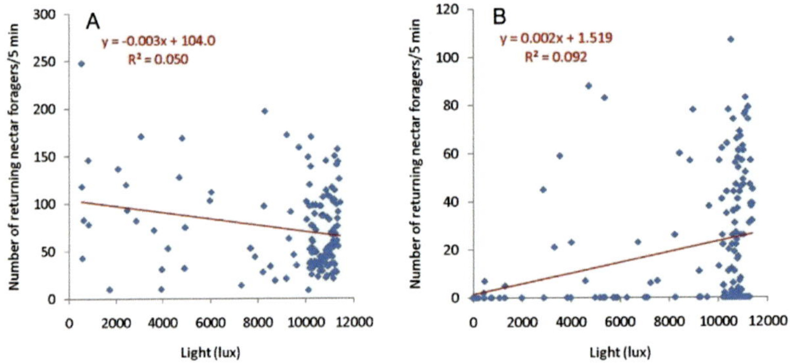

Figure 8.9. Scatter plots showing the relationship between light intensity and the returning nectar foragers. (A) During hot summer, (B) during cold winter.

2. Resource availability

Bees graze mostly for nectar and pollen grains. Adult bees rely on nectar for energy, whilst larvae require pollen for development. Aside from these floral supplies, stingless bees get nesting materials (latex and resin) from plants. The availability of these materials in the environment determines where and how extensively bees forage. Flowering plants with readily available nectar and pollen attract more foraging bees. Adequate resource availability is critical to the health and development of bee colonies. In extreme circumstances, inadequate resources can cause weakened colonies, reduced brood production, and even colony collapse. Bees communicate resource availability through waggle dances and pheromones. When resources are abundant, more bees may be attracted to forage in certain areas. More bees may be recruited to forage at specific locations when resources are abundant. In contrast, competition for limited resources can result in bee disputes or shifts in foraging preferences. Resource availability influences foraging activity in the following ways:

Energy acquisition: Animals forage to get energy from food. When resources are plentiful, foraging may be more efficient, taking less time and effort to meet energy requirements. This can result in shorter foraging periods and more time available for other activities such as reproduction or territory defence.

Time and effort: Scarce resources require more time and effort to locate and exploit. Animals may need to travel longer distances, spend more time searching, or employ more intricate foraging techniques. This can increase energy consumption and impact overall fitness.

Behavioural decisions: The availability of resources determines how an animal makes decisions. For example, when resources are abundant, animals may discriminate more in their foraging decisions, choosing higher quality or favoured meals. In contrast, they may broaden their diet amid scarcity or accept lower-quality resources.

Competition: Resource availability affects competition among individuals and species. In times of scarcity, competition for limited resources intensifies, leading to changes in foraging behaviour such as increased aggression, territoriality, or shifting to alternative food sources.

Reproductive success: Adequate resource availability is crucial for reproductive success. Animals need sufficient energy reserves to invest in offspring, including gestation, lactation, and care. Poor resource availability can lead to reduced reproductive output or even failure.

Population dynamics: Long-term changes in resource availability can impact population dynamics. Periods of resource abundance can support larger populations, while scarcity may lead to population declines or migrations in search of better conditions.

Adaptation and evolution: Over time, foraging strategies and behaviours evolve in response to resource availability. Natural selection favours individuals with behaviours that maximise energy intake relative to available resources, shaping foraging efficiency and adaptability.

3. Colony demands

Colony demands, and bees' foraging activity are tightly interconnected, with bees' foraging behaviour influenced by the colony's needs.

Nutritional needs: Bees forage primarily to collect nectar and pollen, which provide energy for the hive. The colony's size and activity level directly affect the resources required. Larger colonies with more adult bees and brood (developing bees) have higher energy requirements, necessitating more foragers to collect more nectar and pollen.

Brood rearing: The requirement for pollen is very high during brood raising. Nurse bees feed developing larvae with pollen, which is critical to the colony's growth and health. When brood production rises, additional foragers are recruited to collect pollen to meet nutritional demands.

Honey storage: Bees also forage for nectar, which is turned into honey and stored as a long-term energy source. Seasonal fluctuations, meteorological conditions, and the availability of floral resources determine the colony's honey storage requirements. During periods of ample nectar flow, bees increase their foraging efforts to acquire and store excess honey.

Seasonal changes: Colony demands vary periodically, altering foraging activities. For example, during the spring and summer, when floral resources are plentiful and brood rearing is at its optimum, bees increase their foraging efforts to fulfil the colony's growing needs. In contrast, during the winter, when resources are sparse or unavailable, bees may reduce foraging activity and rely on stored honey to support the colony.

Division of labour: Bee colonies have a sophisticated division of labour, with distinct age groups of bees (young workers, middle-aged foragers, senior guardians, etc.) doing specific jobs based on the colony's requirements. Colony demands influence bees' transition from nursing to foraging jobs. Increased demand for nectar and pollen may hasten the development of foraging behaviour in juvenile bees.

Communication and recruitment: Bees convey information about colony demands using various methods, including pheromones and the waggle dance. For example, when resources are plentiful or specific types of pollen are required for brood rearing, bees recruit more foragers to target those resources efficiently.

Resource allocation: Colony demands influence how bees distribute their foraging efforts among various floral resources. Bees may prioritise certain plants or flower kinds that provide vital nutrients (for example, pollens rich in essential proteins) or produce abundant nectar for honey production.

4. Daytime-wise activity

The foraging behaviour of stingless bees is greatly affected by daytime fluctuations, mainly because of the variations in temperature, light intensity, and the abundance of floral resources throughout the day. Stingless bees are active during the day, and their foraging habits are closely linked to the sun's cycle. In the morning, when the sun rises and the temperatures rise, the bees become active and leave the hive to seek out nectar and pollen. In the morning, the bees use the sun's warmth to reach the right body temperature for efficient flight and activity. Throughout the day, the light gets stronger, and bees keep foraging energetically, making the most of the peak blooming times of many flowers that coincide with daylight to maximise pollination. On most plants, the peak foraging activity of Indian stingless bees is 10.00-12.00 h; this period often represents the peak of foraging activity because of the perfect balance of light and temperature. This helps the bees navigate and find flowers more quickly and ensures the highest nectar production in many plant species. If temperatures rise too much, usually in the early afternoon, bees might decrease their activity to prevent overheating and dehydration. They may seek shade or return to the hive to lower their body temperature. In the late afternoon, as the temperature and light intensity decrease, foraging activity slowly decreases, and bees return to the hive to avoid the cooler temperatures and reduce light levels in the evening. Evening signals a reduction in foraging activity for bees, as their navigation and foraging efficiency heavily depend on daylight. This daily rhythm allows bees to make the most of their foraging efforts while reducing the impact of temperature extremes and low light. Knowledge of these daytime patterns is vital for beekeepers and farmers to enhance hive management and crop pollination, promoting bee colonies' well-being and pollinated plants' productivity.

5. Season-wise activity

Weather conditions vary in different seasons, with hot summers, rainy monsoons, and cold winters. Bees' foraging behaviour is greatly influenced by the changing seasons, mainly due to fluctuations in temperature, daylight, and floral resources. When spring arrives, stingless bees leave their winter hibernation and are greeted by a landscape teeming with blossoming flowers that offer abundant nectar and pollen. This abundance is crucial as colonies must replenish their strength and food stores. As temperatures soar in summer, foraging activity peaks and bees utilise extended daylight hours to enhance their gathering efforts. During this time, the variety of flowering plants is at its peak, providing a wide range of foraging opportunities. As bee activity increases, so does the competition and stress on bee populations. As fall approaches, the availability of floral resources diminishes, causing bees to prioritise collecting and storing enough resources for winter survival. Cooler temperatures and shorter days indicate a decrease in foraging. During winter, the foraging activity of stingless bees is very low as bees enter a semi-dormant state, relying on stored honey and pollen to sustain them through the cold months. Therefore, the seasonal cycle determines the pattern of foraging behaviour, impacting stingless bee colonies' survival and well-being.

Foraging patterns

Stingless bees forage on many flowers on a foraging trip, except very large flowers with plenty of floral resources. Floral visitors have evolved various foraging techniques to maximise collecting efficiency, influencing the reproductive success and population structure of the plants they visit (Goulson 1999). A typical foraging sequence includes search, encounter, selection, pursuit, handling, ingestion, and digesting (Bartumeus, 2007).

Several models and assumptions have been developed to optimise animal foraging behaviours. Some of these optimal foraging models are the marginal value theorem (Charnov 1976), ideal free distribution (Fretwell and Lucas 1970), and levy flights (Bartumeus 2007; Sims et al. 2007). The marginal value theorem states that a forager should leave a patch and begin looking for another food source when the rate of food consumption in the patch declines. According to the ideal free distribution model, the evolutionary stable strategy for foragers exploiting a patchy resource is to match the proportion of foragers in each patch to the rate of reward production. Non-random spatial searching, traplining along defined paths, and methodical searching can result in a perfect free distribution. Levy flight patterns, on the other hand, are defined by sequences of randomly oriented straight-line movements with lengths distributed according to an inverse-square power law (Pyke 2015). Levy flights are considered scale-free since the observational scale does not affect their statistical qualities (Reynolds et al. 2007). The floral visitation patterns of stingless bees may be of a random search (levy flights), systematic search (ideal free distribution), or other search strategies. The Indian stingless bee *Tetragonula iridipennis* performs levy flights on most plant species (Figure 8.10). These short-distance flights imply that throughout one complete cycle of foraging trip, the bees stay on flowers most of the time rather than relocating to another distant flower. The path of the flights was indicated by a series of straight-line movements between flowers, with significant changes in directions occurring. The bees traversed the flower area by flying long and short distances in various directions. The consecutive flight movements of stingless bees while foraging were unpredictable and did not follow specific or systematic patterns. Scale-free foraging flights demonstrated by stingless bees display characteristic levy flight patterns, which are an optimal search strategy. The bees randomly explored the flower patch and moved in any direction.

If small-sized flowers cluster on an inflorescence axis, they forage on many inflorescence flowers. A few flowers of the inflorescence receive more than one visit by the individual foragers on a single bout. The distance between two successively visited flowers varies depending on the floral arrangement of the plant species. The distance was very low for closely situated flowers (e.g., fennel) (1–7 cm, 3.80 ± 1.94 cm). The flight distance was also low for close-situated small flowers (e.g., radish) (3–30 cm, 10.57 ± 6.99 cm). They utilise flowers of different levels of height. Though they mostly visit on the same level flowers.

Depending on floral characteristics and the foraging behaviour of the bee species, a flower can receive single or multiple successive visits from a forager. Indian stingless bees mostly perform a single visit on a flower at a time. However, flowers with rich floral resources (e.g., *Solanum sisymbriifolium*) may receive many successive visits by the stingless bees. In the early morning, *Tetragonula iridipennis* visits a flower 6–10 times successively. The interval between consecutive visits to a flower is about 3.50 seconds.

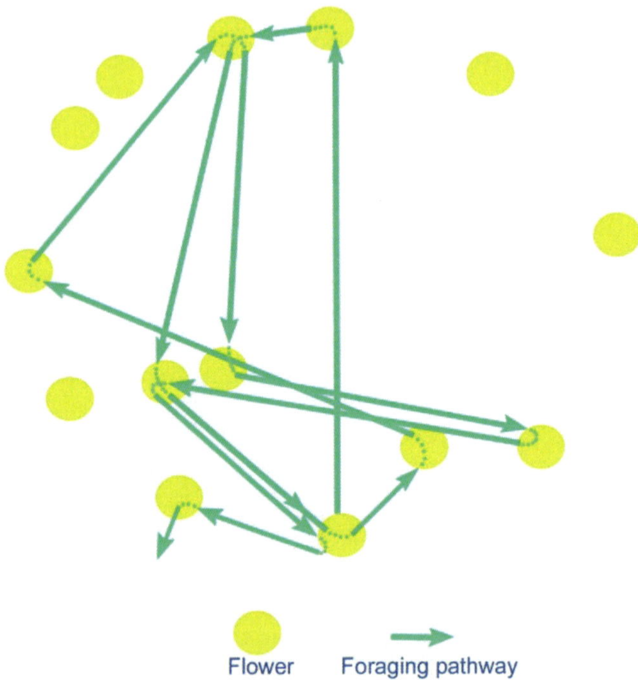

Figure 8.10. The levy flight pattern of Indian stingless bees.

Foraging trip

A foraging trip or foraging bout is when a bee leaves its hive to search for and collect food resources, such as nectar, pollen, water, or propolis, and then returns to the hive. This trip can vary in duration and distance depending on several factors, including the availability of resources, weather conditions, and the bee's species and role within the colony. During a foraging bout, a bee may visit multiple flowers and locations before returning to the hive to deposit the collected resources. The efficiency and success of these foraging trips are crucial for the survival and productivity of the bee colony.

Daily number of foraging trips

An individual forager did not uniformly forage throughout the day time. Initially, they foraged at regular intervals. Later on, their foraging trip became irregular. Throughout the day, the average number of foraging trips for a nectar forager was 10.93 ± 4.77; for a pollen forager, it was 8.08 ± 3.53; for a resin forager, it was 5.70 ± 3.06; and for a latex forager, it was 6.36 ± 3.39 (Layek et al. 2024). The number of foraging trips significantly differed with seasons and varied among different studied colonies. Seasonally, the highest number of trips was recorded during summer, followed by spring, and the lowest was recorded during the monsoon season (Table 8.1). When honey was made available at the nest entrance to estimate maximum potential, the number of trips performed by an individual was much higher (188.90 ± 25.77 trips/day) than the actual values found in nature. As the distance of food source increases from the nest, the number of trips gradually decreases.

Table 8.1. The mean number of foraging trips performed by a worker daily (Layek et al. 2024).

Parameter	Colony A	Colony B	Colony C	Colony D	Mean ± S. D.
Summer					
Nectar forager	11.90	13.90	19.70	17.40	15.72^{a} ± 4.43
Pollen forager	8.30	9.80	13.30	12.80	11.05^{a} ± 3.47
Resin forager	7.00	7.70	8.60	5.10	7.10^{ab} ± 2.57
Latex forager	8.10	8.80	9.60	5.80	8.08^{a} ± 3.05
Monsoon					
Nectar forager	5.50	6.50	7.20	6.80	6.50^{e} ± 3.15
Pollen forager	4.10	4.80	5.60	4.90	4.85^{d} ± 2.80
Resin forager	2.10	2.80	3.20	1.20	2.32^{d} ± 1.87
Latex forager	2.60	3.10	3.50	1.60	2.70^{d} ± 2.14
Autumn					
Nectar forager	6.30	8.70	10.40	9.30	8.68^{d} ± 2.87
Pollen forager	4.90	6.60	7.80	7.30	6.65^{c} ± 2.51
Resin forager	5.70	6.90	7.10	3.70	5.85^{c} ± 2.68

Latex forager	6.60	7.60	7.80	4.10	6.52^{bc} ± 3.14
Late autumn					
Nectar forager	9.10	10.70	11.30	10.70	10.45^{c} ± 3.00
Pollen forager	6.80	7.80	8.40	8.20	7.80^{bc} ± 2.28
Resin forager	8.40	8.90	9.20	3.90	7.60^{a} ± 2.96
Latex forager	8.80	9.70	9.80	4.50	8.20^{a} ± 3.28
Winter					
Nectar forager	7.90	10.80	12.70	11.70	10.78^{c} ± 3.95
Pollen forager	6.40	8.20	9.40	9.00	
Resin forager	5.30	6.00	6.80	2.40	5.12^{c} ± 2.62
Latex forager	5.70	6.60	7.40	2.80	5.62^{c} ± 2.98
Spring					
Nectar forager	10.60	12.20	16.20	14.90	13.48^{b} ± 4.57
Pollen forager	7.70	9.20	11.40	11.10	9.85^{a} ± 3.48
Resin forager	6.40	6.70	7.10	4.50	6.18^{bc} ± 2.55
Latex forager	7.70	7.40	8.10	4.90	7.02^{ab} ± 2.40
Mean ± S. D.					
Nectar forager	8.55 ± 3.74	10.47 ± 3.96	12.92 ± 5.09	11.80 ± 5.06	
Pollen forager	6.37 ± 2.82	7.73 ± 3.11	9.32 ± 3.55	8.88 ± 3.85	
Resin forager	5.82 ± 2.77	6.50 ± 3.03	7.00 ± 3.04	3.47 ± 2.10	
Latex forager	6.58 ± 3.16	7.20 ± 3.43	7.70 ± 3.31	3.95 ± 2.30	

Note. Means in the column followed by the same alphabets do not differ significantly by DMRT (at 5%), S. D.- standard deviation.

Duration of a foraging trips

The foraging trip duration showed significant differences between nectar, pollen, resin, and latex foragers ($F3, 236 = 37.54$, $P = 7.12E\text{-}20$). Nectar foragers had the shortest average duration for a foraging trip (10.75 ± 4.30 minutes), followed by latex foragers (14.05 ± 4.83 minutes), resin foragers (16.00 ± 6.43 minutes), and pollen foragers (26.67 ± 14.82 minutes). Furthermore, the duration of foraging trips varied throughout the day, with the trip duration being shortest in the morning and gradually increasing over time (Figure 8.11).

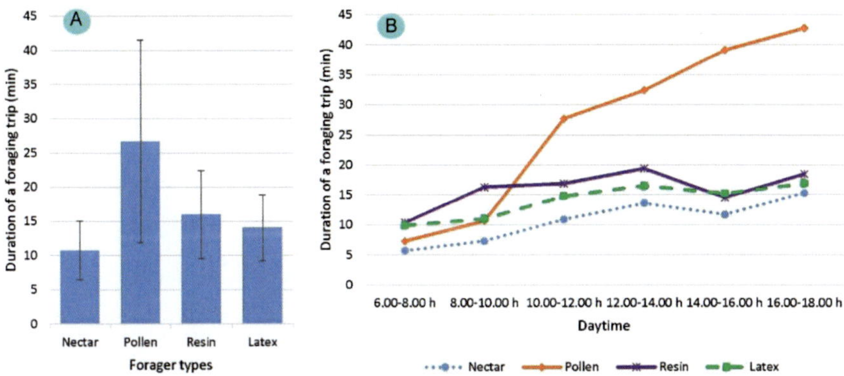

Figure 8.11. Duration of a forage trip for different types of foragers. (A) The average duration of a foraging trip considering throughout the daytime, (B) duration of a foraging trip at different day times.

Factors influence the number of foraging trip

Several factors influence the number of foraging trips bees make, affecting their efficiency and the overall productivity of the colony. Key factors include:

- **Availability of floral resources:** The density and diversity of flowering plants directly impact the number of foraging trips. Bees can make fewer trips to meet their nutritional needs when flowers are abundant and rich in nectar and pollen.
- **Distance to food sources:** The proximity of floral resources to the hive affects foraging activity. Closer food sources mean shorter

trips and more frequent foraging, while distant resources require more energy and time, reducing the number of trips.

- **Weather conditions:** Temperature, humidity, and precipitation significantly affect foraging behaviour. Optimal temperatures encourage frequent foraging, while extreme heat, cold, or rain can reduce the number of trips as bees avoid adverse conditions.
- **Time of day:** Bees are diurnal and typically forage during daylight hours. The length of the day influences how many trips they can make, with longer days allowing for more foraging opportunities.
- **Colony health:** The bee colony's health plays a crucial role. Healthy colonies with a strong population can sustain more foragers, increasing the number of trips. Diseases, pests, or pesticide exposure can weaken bees, reducing their foraging efficiency and frequency.
- **Colony needs:** The colony's internal requirements, such as feeding larvae or building up food reserves for winter, drive the frequency of foraging trips. High demand within the hive can increase foraging activity.
- **Experience of foragers:** The age and experience of the forager bees can also play a role. Experienced foragers are more efficient and can complete more trips than younger, inexperienced bees.
- **Predation and competition:** Predators or competitors can affect foraging trips. Bees may reduce their activity in areas with high predator presence or intense competition from other pollinators.

Resource carrying capacity

On a trip, a nectar forager collected about 1.20 mg of nectar from local vegetation. The maximum nectar-carrying potential (1.50 mg/trip) estimated by the supplementary honey-feeding method was slightly higher than the value obtained in nature. The bee species carried different amounts of corbicular loads, viz. pollen, resin, and latex. On average, each pollen forager carried about 3.75 mg of pollen loads, the resin forager carried 1.20 mg of resin loads, and the latex forager carried about 2.26 mg of latex loads per trip.

Factors influencing the carrying capacity of stingless bees

Several factors influence the resource-carrying capacity of bees, determining how much nectar, pollen, and other resources they can transport back to the hive during foraging trips. These factors include:

- **Bee species:** Different bee species have varying carrying capacities. For example, honeybees can carry more resources than smaller bee species due to their larger body size and specialised structures like pollen baskets on their hind legs.
- **Individual bee size and health:** Individual bees' physical size and health influence their carrying capacity. Larger, healthy bees can carry more nectar and pollen compared to smaller or unhealthy ones. Nutritional deficiencies or diseases can weaken bees, reducing their ability to transport resources efficiently.
- **Flight muscle strength and efficiency:** The strength and efficiency of a bee's flight muscles impact how much weight it can carry. Stronger muscles enable bees to carry heavier loads, while efficient muscle function allows for longer foraging trips without exhaustion.
- **Experience and age of foragers:** Experienced foragers are generally more efficient in resource collection and transport than younger, inexperienced bees. Age and experience can influence a bee's ability to optimise its carrying capacity during foraging trips.
- **Resource quality and quantity:** The quality and concentration of nectar and pollen available in flowers affect how much bees can carry. High-quality, concentrated nectar allows bees to gather more energy in a single trip, while nutrient-rich pollen can be packed more densely.
- **Environmental conditions:** Weather conditions such as temperature, humidity, and wind speed can influence carrying capacity. Optimal temperatures and humidity levels enhance foraging efficiency, while extreme conditions can hinder a bee's ability to carry and transport resources.
- **Foraging distance:** The distance between the hive and foraging sites affects carrying capacity. Longer distances require more energy, reducing the resources bees can carry back to the hive.
- **Floral characteristics:** Flowers' structure and accessibility impact how efficiently bees can collect nectar and pollen. Flowers with easily accessible resources allow bees to gather more quickly and efficiently, maximising their carrying capacity.

Resource loading and unloading time

The minimum nectar loading time estimated by the artificial honey feeding method was 47.96 ± 3.64 seconds. The loading time does not depend on the distance of the resource. However, it can greatly vary among different flowering species. In *Brassica nigra* (medium-sized flower with more nectar), the average loading time was 4.47 ± 1.14 minutes. The loading time was lesser in the morning (3.70 ± 0.82 minutes), thereafter increased in noon (4.50 ± 1.08 minutes) and afternoon (5.20 ± 1.03 minutes). In *Coriandrum sativum* (small-sized flower with less nectar), the loading time was significantly higher (18.17 ± 4.93) in comparison to the above plant species.

Pollen loading time also varied according to plant species as well as the time of collection. The pollen loading times on *Luffa aegyptiaca* and *Turnera ulmifolia* (large-sized flowers with huge pollen grains in exposed stamens) were 10.13 ± 5.44 minutes and 8.67 ± 2.87 minutes, respectively. On *Solanum sisymbriifolium*, the loading time was much higher (32.80 ± 10.82 minutes) in comparison to the above species. In all cases, the pollen loading time was significantly lower during the morning. As time passed, pollen availability decreased, resulting in increased pollen loading time.

The minimum nectar unloading time estimated by the artificial honey feeding method was 82.74 ± 13.73 seconds. It does not vary according to daytime. However, unloading time was slightly varied among different distances of the resource. In natural habitats, resource unloading time significantly differed among the resource types. The resource unloading time was lowest for pollen load (77.17 ± 18.33 seconds), followed by nectar load (83.78 ± 22.29 seconds), resin load (307.86 ± 91.07 seconds), and latex load (392.93 ± 80.70 seconds).

The time bees load and unload resources, such as nectar and pollen, is influenced by several factors that impact their overall foraging efficiency. The floral morphology plays a significant role; flowers with complex structures or deep nectaries require more time for bees to access and extract resources. The concentration and quality of nectar also affect loading time, as more diluted nectar takes longer to collect. Environmental conditions like temperature and humidity influence the viscosity of nectar and bees' agility, affecting how quickly they can gather and transport resources. Within the hive, the efficiency of resource transfer depends on the availability and readiness of worker bees to receive and process the incoming nectar and pollen. The health and experience of the forager bees are crucial, as healthy, experienced bees can optimise their collection and unloading processes. Additionally, the hive's internal needs and communication through the waggle dance can dictate how urgently resources need to be processed, influencing the pace of unloading. Predation risk and competition from other pollinators can also impact how quickly bees complete their foraging trips, as they may need to hurry to avoid threats or outcompete others for floral resources. Understanding these factors helps assess bee colonies' overall foraging efficiency and productivity.

Flight distance of Indian stingless bees

The flight range of bees can be determined by artificial feeder methods and capture-recapture methods. The foraging distance for a bee species is a vital criterion for the suitability of its management and application in agricultural fields. The flight distance of bees depends on their body size; larger bees have a greater flight range (Araújo et al. 2004; Greenleaf et al. 2007). The Indian stingless bees have a flight range of about 900 metres (Layek et al. 2024). The flight range of the foragers slightly varies among different colonies and is influenced by external and internal factors.

Factors affecting the foraging range of stingless bees

- **Resources availability:**

The distribution and availability of floral resources are critical determinants of the flight range of bees, directly impacting their foraging efficiency and energy expenditure. When floral resources are abundant and evenly distributed within a habitat, bees can fulfil their nutritional needs within a limited area, reducing their travel distance. However, in environments where flowers are sparse or clustered in isolated patches, bees must fly longer distances to find adequate food, thereby increasing their foraging range. This extended travel demands more energy and increases the risks associated with longer flights, such as predation and adverse weather exposure. Additionally, seasonal variations in floral availability can cause fluctuations in flight range, with bees travelling further during periods of resource scarcity. The spatial and temporal distribution of flowers thus plays a vital role in shaping bee foraging behaviour, influencing their ecological roles and the overall health of the colony. If surrounding floral resources are available, they travel less distance. During monsoon and autumn (the dearth period for bees), stingless bees expanded their foraging radius (Layek et al. 2024).

- **Habitat type:**

Habitat plays a crucial role in determining the flight range of bees by affecting the availability and distribution of floral resources, nesting sites, and overall environmental conditions. In diverse and resource-rich habitats, such as meadows or mixed agricultural landscapes, bees can find ample nectar and pollen within shorter distances, minimising the need for extensive travel. Conversely, bees are forced to extend their foraging range to locate sufficient resources in degraded or fragmented habitats, where flowers are scarce and patchily distributed. Urban environments can also influence flight range, as bees may need to navigate through or around buildings and other obstacles, potentially increasing travel distances. Additionally, the presence of natural features like forests, water bodies, and topographical variations can either facilitate or hinder bee movement, further shaping their foraging patterns. Thus, the quality and configuration of the habitat directly impact the energy expenditure, foraging efficiency, and overall survival of bee populations.

- **Competition with other visitors:**

Competition can significantly influence the flight range of bees by driving them to explore larger areas in search of food. When multiple bees or other pollinators vie for the same floral resources, local supplies can quickly be depleted, necessitating longer foraging trips. In areas with high competition, bees might need to extend their flight range to locate adequate nectar and pollen sources that are not already exploited. This increased competition can occur within the same species (intraspecific) and between different species (interspecific), intensifying the pressure on available resources. As a result, bees may adopt strategies to outcompete others, such as foraging at different times of the day or targeting less competitive floral patches, even if these are farther from the hive. The need to travel greater distances due to competition demands more energy and time and increases the risks associated with foraging, such as predation and exposure to adverse weather. Thus, competition is crucial in shaping bee populations' flight range and foraging dynamics.

- **Weather conditions:**

Weather conditions profoundly impact the foraging range of bees, influencing their ability to search for food and return safely to the hive. Temperature, humidity, wind, and precipitation are critical factors that affect bee foraging behaviour. Optimal temperatures, typically between 20°C to 30°C (68°F to 86°F), enable bees to fly efficiently and cover greater distances. Extreme heat or cold can reduce their foraging range, as bees may become lethargic or risk hypothermia. High humidity can hinder flight by making wings heavy with moisture, while strong winds can physically impede bees' ability to navigate and increase their energy expenditure, limiting their range. Rainy conditions are particularly detrimental, as bees avoid foraging in wet weather to prevent wing damage and maintain body temperature. Favourable weather conditions promote extended foraging ranges, while adverse conditions constrain bees from foraging closer to their hives.

- **Seasonal impact:**

Seasons significantly influence bees' flight range by altering floral resource availability and environmental conditions throughout the year. In spring and summer, when flowers are abundant and weather conditions are generally favourable, bees can find ample nectar and pollen within shorter distances, reducing flight range. During these seasons, the warm temperatures and longer daylight hours facilitate extended foraging activities. Conversely, floral resources become scarce in autumn and winter as many plants stop blooming and cooler temperatures prevail. As a result, bees must travel further to locate sufficient food, increasing their flight range. Additionally, adverse weather conditions such as rain, wind, and cold during these seasons can further restrict foraging, forcing bees to remain closer to the hive or cease foraging altogether. Seasonal changes thus play a crucial role in determining the flight range and foraging behaviour of bees, impacting their ability to gather resources and sustain the colony.

CHAPTER 9

PLANT RESOURCES FOR INDIAN STINGLESS BEES

Different types of plant resources

For the colony's construction, sustenance, and growth, stingless bees collected nesting materials, food rewards, and water (Figure 9.1). As nesting materials, they collect latex, resin, plant fragments, and sometimes soil particles (Layek et al. 2023b). Food rewards mainly came from nectar and pollen of flowers. Besides floral rewards, they also collect sugary saps from fruits and sweets. These different types of plant resources they collect from diverse plant species. Therefore, the diversity and abundance of flowering plants directly impact stingless bee populations, as a continuous supply of various blooms ensures that bees have access to necessary resources throughout the year. In turn, during floral rewards collections, they support pollination, which is critical for ecosystem stability and agricultural productivity.

Understanding nectar, pollen, and resin sources for bees is crucial for several interrelated reasons that impact bee health, ecological balance, and agricultural productivity. Resin, collected for hive construction and protection, contributes to the hive's hygiene and defence mechanisms. Nectar provides essential carbohydrates and is a primary energy source for bees, crucial for their survival and hive maintenance. Rich in proteins and lipids, pollen is vital for brood development and overall colony growth. Knowledge of these resources enables beekeepers to manage hives more effectively, ensuring bees have access to a balanced diet supporting their health and productivity. It also aids in habitat conservation by emphasising the importance of preserving diverse flowering plants and resin-producing trees, which are critical for maintaining healthy bee populations and promoting biodiversity. Furthermore, this understanding informs agricultural practices by encouraging the cultivation of pollinator-friendly crops and minimising pesticide use, thereby supporting bee health, crop yields, and ecosystem stability. By recognising the importance of these resources, we

Figure 9.1. (A) Returning foragers with resources, (B–C) foragers with nectar load in the crop, (D–E) foragers with corbiclular pollen loads, (F–G) foragers with corbicular latex loads, and (H–I) foragers with corbicular resin loads. Scale bars = 5 mm.

can develop strategies to protect and enhance bee habitats, ensuring the resilience and sustainability of both natural and agricultural environments and ultimately fostering a thriving ecosystem where bees and other pollinators can flourish.

Latex and resin sources for stingless bees

Latex is a milky fluid composed of various organic compounds in certain plants' latex vessels or ducts, particularly within the bark, leaves, and stems. Latex consists of water, proteins, alkaloids, starches, sugars, oils, tannins, resins, and gums. Resin is typically a sticky, viscous liquid that can harden upon exposure to air. It is usually amber in colour, though it can vary depending on the plant species. Resin is composed mainly of terpenes and resin acids, volatile organic compounds. It may also contain essential oils, phenolic compounds, and other substances.

Stingless bees collect plant latex and resin and mix them with bee wax and other substances to make these nesting materials (i.e., cerumen and propolis). Cerumen is mainly utilised to construct the nest components (including food pots, brood cells, involucrum, and other supporting strands). In contrast, propolis is used as sealing material and for protective purposes; it is found in the joining point of the top cover sheet of a managed hive. Cerumen and propolis provide antiseptic nest environments as these substances have antimicrobial activity (Park et al. 1998).

The plants utilised as latex and resin sources for bees can be determined by field surveys, chemical analysis of bee-collected latex/resin, or pollen content of bee-collected latex and resin. Some researchers analyse bee-collected latex and resin pollen content to determine the botanical origins. However, the pollen content of corbicular and in-hive stored latex/resin, cerumen, and propolis does not truly infer its botanical origin (Layek et al. 2023b). Therefore, latex and resin sources are determined by techniques like direct observation or chemical profiling.

Layek et al. (2023b) recorded a few plants as latex/resin sources for the stingless bee *Tetragonula iridipennis* in West Bengal (Table 9.1). Important (i.e., frequently observed) resin/latex sources were *Artocarpous heterophyllus*, *Calotropis gigantea*, *Ficus benghalensis*, *Ficus religiosa*, *Mangifera indica*, *Tabernaemontana divaricata*, and *Vachellia nilotica* (Layek et al. 2023b). Highly represented plant families were Anacardiaceae, Apocynaceae, Fabaceae, and Moraceae.

Table 9.1. Latex and resin sources of Indian stingless bees (*Tetragonula iridipennis*) in West Bengal, India.

Family	Plant species	Relative frequency (%)	Plant organ	Resource
Anacardiaceae	*Lannea coromandelica* (Houtt.) Merr.	2.92	bark	resin
	Mangifera indica L.	18.99	bark	resin
	Spondias pinnata (L. f.) Kurz	2.19	bark	resin
Apocynaceae	*Alstonia scholaris* (L.) R. Br.	2.92	bark	latex
	Calotropis gigantea (L.) Dryand	3.65	leaf, fruit	latex
	Cascabela thevetia (L.) Lippold	2.19	leaf, fruit	latex
	Catharanthus roseus (L.) G. Don	1.46	immature leaf	latex
	Tabernaemontana divaricata R. Br. ex Roem. & Schult	4.38	Flower bud, leaf petiole	latex
Caricaceae	*Carica papaya* L.	1.46	young flower bud, fruit	latex
Cleomaceae	*Cleome viscosa* L.	0.73	immature leaf	resin
Dipterocarpaceae	*Shorea robusta* Roth	2.19	bark	resin
Euphorbiaceae	*Jatropha curcas* L.	2.19	immature leaf	resin
	Jatropha gossypiifolia L.	0.73	immature leaf	resin
Fabaceae	*Albizia lebbeck* (L.) Benth.	2.19	bark	resin
	Butea monosperma (Lam.) Taub.	8.03	bark	resin
	Senna occidentalis (L.) Link	1.46	immature leaf	resin
	Vachellia nilotica (L.) P. J. H. Hurter & Mabb.	5.11	bark	resin
Malvaceae	*Bombax ceiba* L.	0.73	bark	resin
	Ceiba pentandra (L.) Gaertn.	2.92	bark	resin

Moraceae	*Artocarpus heterophyllus* Lam.	6.57	bark, immature fruit	latex
	Ficus benghalensis L.	10.22	bark	latex
	Ficus religiosa L.	5.84	bark	latex
Moringaceae	*Moringa oleifera* Lam.	8.03	bark	resin
Papaveraceae	*Argemone maxicana* L.	1.46	stem	latex
Sapotaceae	*Mimusops elengi* L.	1.46	bark, immature leaf	latex

Stingless bees select specific plant parts for latex and resin collection based on a nuanced interplay of factors. They target areas where the latex or resin is easily accessible, such as exposed bark or active resin ducts on leaves. The physical properties of the resin, including its viscosity and adhesive qualities, play a crucial role; bees prefer substances that are neither too thick nor too thin to facilitate efficient collection and transport. Additionally, bees are guided by the chemical composition of the resin, seeking out those with beneficial nutrients and optimal flow rates. Their choices are influenced by past experiences, where they learn to favour plants and parts that consistently provide high-quality resin. Environmental conditions, such as temperature and humidity, further affect their selection, as these factors influence the secretion and consistency of the latex or resin. Overall, bees integrate these criteria to maximise their foraging efficiency and ensure the collected resin meets the needs of their hive. The bee species collected latex/resin from different plant parts, including injured bark of stem, young leaves, and immature fruits (Figure 9.2).

Characteristics of latex and resin-providing plants

The latex and resin-providing plants are primarily of tree habit (76%), following shrubs (16%) and herbs (8%). Many trees have punctured bark secreting latex or resin, and stingless bees collect them. The available amount of secreted latex or resin is also higher in trees than in herbs, shrubs, and climbers. Additionally, trees provide more security and less disturbance to the foragers during resource collection than other plant habits. The bark characteristics do not influence their latex and resin collection. Latex and resin-providing plants have both hard and soft, rough or smooth textures.

Figure 9.2. Stingless bees collect latex from jackfruit (A) and resin from damaged mango bark (*Mangifera indica*) (B). Scale bars = 10 mm.

Nectar and pollen sources for stingless bees

Nectar is a sugary liquid secreted by plants, primarily by specialised glands called nectaries, located in various parts of the plant, such as flowers, stems, or leaves. Nectar is composed mainly of water and sugars, primarily fructose, glucose, and sucrose. It may also contain small amounts of amino acids, vitamins, minerals, and other compounds, contributing to its flavour and nutritional value. The composition of nectar can vary widely among plant species, influencing its sweetness and viscosity. Nectar is fundamental to bee colonies' nutritional well-being and functional stability, supporting their energy needs, reproductive success, and hive sustainability.

Pollen is the male gametophyte of gymnosperms and angiosperms. Pollen is produced in the anthers, which are part of the stamen, the male reproductive organ of a flower. Pollen is a crucial nutritional resource for bees, providing a rich source of proteins, lipids, vitamins, minerals, and essential amino acids. It serves as the primary protein source for bees, especially for the development of larvae and the overall health of the hive.

A forager bee collects nectar and pollen from flowers through a highly specialised process involving several coordinated steps. Upon locating a flower, the bee uses its compound eyes to detect visual cues such as colour and patterns and its antennae to sense floral scents. Landing or hovering, the bee extends its proboscis, a tubular mouthpart, to access the nectar within the flower's nectaries. Using capillary action and suction, the bee draws nectar into its honey stomach, a specialised organ for temporary storage. Foragers' hairy bodies brush against the anthers, collecting pollen. Bees groom themselves using their front and middle legs, transferring the pollen to specialised structures on their hind legs known as pollen baskets. Once these baskets are full, bees return to the hive, depositing the pollen into storage cells. Other workers may mix the pollen with nectar and enzymes within the hive to produce "bee bread," a nutritious food source for the larvae and young bees.

Determining nectar sources for a bee species involves a multifaceted approach that integrates observational, experimental, and analytical methods to pinpoint which plants provide the most valuable resources. Initially, field observations are conducted to identify which plants bees frequent for nectar collection, often focusing on flowering plants in the bee's habitat. Pollen analysis of honey (i.e., melissopalynological studies) is frequently used to determine the botanical origin of honey, i.e. nectar sources. Bee's body surface and gut pollen content also indicate their nectar sources. Chemical analysis of nectar samples can further help identify plant sources by comparing the chemical signatures found in the nectar to those of known plants. Pollen sources can be determined by conducting field surveys focusing on floral visitors and pollen analysis of bee-collected pollen loads. Environmental factors such as temperature, humidity, and seasonal variations also play a role in determining floral resources (nectar and pollen) availability and quality. Combining these methods with knowledge from local beekeepers and ethnobotanical research can offer a comprehensive understanding of the nectar and pollen sources important to a particular bee species. This holistic approach helps identify the plants that bees rely on, understand the ecological dynamics, and ensure the sustainability of both bee populations and plant-pollinator interactions.

Table 9.2. Nectar and pollen sources for stingless bees in West Bengal (data sources: Layek and Karmakar 2018; Bisui et al. 2019, 2022; authors own observation).

Family	Plant species	Flowering season						Floral resources	
		I	II	III	IV	V	VI	Nectar	Pollen
Acanthaceae	*Barleria cristata* L.	*						-	+
	Hygrophila auriculata Schumach				*	*	*	+	+
	Justicia adhatoda L.	*						+	+
	Justicia simplex D. Don	*	*					+	+
Aizoaceae	*Trianthema portulacastrum* L.	*	*					+	+
Amaranthaceae	*Amaranthus spinosus* L.	*	*					-	+
	Chenopodium album L.	*					*	-	+
	Spinacia oleraea L.					*	*	-	+
Amaryllidaceae	*Allium cepa* L.					*	*	++	+
	Crinum asiaticum L.	*	*					-	+
Anacardiaceae	*Lannea coromandelica* (Houtt.) Merr.						*	+++	+++
	Mangifera indica L.					*		+	-
	Semecarpus anacardium L.f.	*	*					+	+
	Spondias pinnata (L.f.) Kurz						*	++	++
Apiaceae	*Coriandrum sativum* L.					*	*	+++	+++
	Foeniculum vulgare Mill.						*	++	++
Apocynaceae	*Alstonia scholaris* (L.) R.Br.					*		+	+
	Wrightia tinctoria (Roxb) R.Br.	*						+	+
Arecaceae	*Borassus flabellifer* L.	*					*	+++	+++
	Cocos nucifera L.	*	*	*	*	*	*	+++	+++
	Phoenix sylvestris L. Roxb.					*		++	+++
Asparagaceae	*Agave amica* (Medik.) Thiede & Govaerts	*	*					-	++
Asteraceae	*Chrysanthemum indicum* L.					*		-	+

	Eupatorium odoratum L.				*	*		+	+
	Mikania scandens B.L.Rob.				*	*		+	+
	Tagetes erecta L.					*		-	+
	Taraxacum officinale (L.) Weber ex F.H.Wigg.	*					*	+	+
	Tridax procumbens L.	*	*	*	*			+	+
	Xanthium strumarium L.			*	*			-	+
Bignoniaceae	*Millingtonia hortensis* L.f.	*				*		+	+
Brassicaceae	*Brassica juncea* (L.) Czern.					*		+++	+++
	Raphanus sativus L.					*	*	++	+++
Cannabaceae	*Trema orientale* (L.) Blume	*	*	*				-	+++
Capparaceae	*Capparis zeylanica* L.	*					*	-	+
Caricaceae	*Carica papaya* L.	*	*	*	*	*	*	-	+
Cleomaceae	*Cleome viscosa* L.	*	*					+	+
Combretaceae	*Combretum indicum* (L.) DeFilipps	*	*					+	+
	Terminalia arjuna (Roxb.) Wight & Arn.	*						+	+
	Terminalia bellerica (Gaertn.) Roxb.	*						+	+
	Terminalia chebula Retz.	*						+	+
Commelinaceae	*Commelina benghalensis* L.		*	*	*			+	-
	Cyanotis axillaris (L.) D.Don ex Sweet		*	*	*			+	+
	Murdannia nudiflora (L.) Brenan			*	*			-	+
Convolvulaceae	*Evolvulus nummularius* (L.) Linn.	*	*					+	+
Cornaceae	*Alangium salviifolium* (L.f.) Wangerin	*					*	+	++
Cucurbitaceae	*Citrullas lanatus* (Thunb.) Matsum. & Nakai						*	+	+
	Coccinia grandis (L.) Voigt	*	*					+	+
	Cucumis sativus L.	*	*	*			*	+	+
	Luffa acutangula (L.) Roxb.	*	*					+	+
	Luffa aegyptiaca Mill.	*	*	*				+	++
	Momordica charantia L.	*	*				*	++	++
Dipterocarpaceae	*Shorea robusta* Roth	*					*	+	+

Euphorbiaceae	*Chrozophora rottlerin* (Geiseler) A.Juss. ex Spreng.	*	*					-	+
	Croton bonplandianus Baill.	*	*					-	+
	Euphorbia milii Des Moul.	*	*	*	*	*	*	-	+
	Jatropha carcus L.	*	*					+	-
	Jatropha gossypiifolia L.	*	*					+	-
	Ricinus communis L.			*	*	*		-	+
Fabaceae	*Acacia auriculiformis* A.Cunn. ex Benth.		*	*	*			+	+
	Albizia lebbeck (L.) Benth.	*	*				*	+	+
	Bauhinia acuminata L.	*	*					+	+
	Butea monosperma (Lam.) Taub.						*	+	-
	Cajanus cajan (L.) Huth			*	*	*		+	+
	Calliandra haematocephala Hassk.		*					+	+
	Cassia fistula L.	*					*	+	+
	Dalbergia sissoo Roxb.						*	+	+
	Delonix regia (Boj. Ex Hook.) Raf.	*	*					+	+
	Lablab purpureus (L.) Sweet					*		+	-
	Leucaena leucocephala (Lam.) de Wit	*	*					+	+
	Millettia pinnata (L.) Pierre	*						++	+
	Mimosa pudica L.				*	*		+	+
	Peltophorum pterocarpum (DC.) K. Heyne	*	*					+++	+++
	Tamarindus indica L.	*	*					+	+
	Tephrosia purpurea (L.) Pers.		*	*	*	*		+	+
	Vachellia nilotica (L.) P.J.H. Hurter & Mabb.	*	*					+	+
Lamiaceae	*Ocimum basilicum* L.	*	*	*	*	*		+	+
	Ocimum tenuiflorum L.	*	*	*	*	*		+	+
	Salvia splendens Sellow ex J.A. Schultes					*		+	+
	Tectona grandis L.f.		*					+	+
	Vitex negundo L.	*	*					+	-

Linderniaceae	*Lindernia crustacea* (L.) Cham. & Schltdl.		*	*				+	+
Loranthaceae	*Dendrophthoe falcata* (L.f.) Ettingsh	*	*					-	+
Lythraceae	*Lagerstroemia speciosa* (L.) Pers.	*	*					+	+
	Punica granatum L.	*	*			*	*	-	+
Malvaceae	*Bombax ceiba* L.						*	+	+
	Ceiba pentandra (L.) Gaertn.						*	+	+
	Corchorus capsularis L.	*	*					+	+
	Corchorus olitorius L.	*	*					+	+
	Grewia asiatica L.				*	*		+	+
Meliaceae	*Azadirachta indica* A. Juss.	*						+	+
	Melia azedarach L.						*	+	+
	Swietenia mahagoni (L.) Jacq.	*	*					+	+
Menispermaceae	*Tinospora cordifolia* (Thunb.) Miers	*	*					+	+
Moraceae	*Streblus asper* Lour.	*					*	-	+
Moringaceae	*Moringa oleifera* Lam.					*		+	+
Myrtaceae	*Eucalyptus tereticornis* Sm.			*	*	*		+	+
	Psidium guajava L.	*	*				*	-	+
	Syzygium cumini (L.) Skeels.						*	+	+
	Syzygium jambos L. (Alston)					*		+	+
	Syzygium reticulatum (Wight) Walp.	*						+	+
	Syzygium samarengense (Blume) Merr. & L.M. Perry						*	-	+
Nelumbonaceae	*Nelumbo nucifera* Gaertn.	*						-	+
Nictaginaceae	*Boerhavia diffusa* L.	*	*					+	+
Oxalidaceae	*Averrhoa carambola* L.	*	*					+	+
	Oxalis corniculate L.		*	*	*	*		+	+
Papaveraceae	*Papaver somniferum* L.					*		-	+
Passifloraceae	*Passiflora foetida* L.	*	*					+	+

	Turnera ulmifolia L.	*	*	*	*	*	*	-	+
Pedaliaceae	*Sesamum indicum* L.	*						+	+
Phyllanthaceae	*Bridelia retusa* (L.) A. Juss.		*	*				+	+
	Phyllanthus emblica L.	*						+	+
Poaceae	*Brachiaria mutica*	*	*					-	+
	Zea mays L.	*	*				*	-	+++
Polygonaceae	*Antigonon leptopus* Hook. & Arn.	*	*	*				+	-
Portulacaceae	*Portulaca oleracea* L.	*	*					+	+
Ranunculaceae	*Nigella sativa* L.						*	+	+
Rhamnaceae	*Ziziphus mauritiana* Lam.			*	*			+	+
	Ziziphus oenopila (L.) Mill.			*	*			+	+
Rosaceae	*Rosa chinensis* Jacq.	*	*	*	*			-	+
Rubiaceae	*Haldina cordifolia* (Roxb.) Brandis		*					-	+
	Ixora arborea Roxb. Ex Sm.	*	*					-	+
	Meyna spinosa Roxb.ex Link	*						+	+
	Morinda citrifolia L.	*						-	+
	Neolamarckia cadamba (Roxb.) Bosser		*					-	+
Rutaceae	*Aegle marmelos* (L.) Correa	*						+	+
	Bergera koenigii L.						*	+	+
	Citrus x aurantiifolia (Christm.) Swingle	*	*				*	+	+
	Citrus maxima (Burm.) Merr.						*	+	+
	Murraya paniculata (L.) Jack	*	*					+	+
Salicaceae	*Flacourtia jangomas* (Lour.) Raeusch.						*	+	+
Sapindaceae	*Litchi chinensis* Sonn.						*	+	+
	Sapindus mukorossi Gaertn.					*		+	+
Sapotaceae	*Madhuca longifolia* (J. Konig) J.F. Macbr.						*	+	+

	Mimusops elengi L.	*						+	+
Simaroubaceae	*Ailanthus excelsa* Roxb.						*	-	+
Solanaceae	*Capsicum frutescens* L.	*	*	*	*	*	*	+	+
	Petunia × alkinsiana (Sweet) D. Don ex W.H. Baxter					*		+	+
	Solanum melongena L.	*	*				*	-	+
	Solanum sisymbriifolium Lam.	*	*				*	-	+
Ulmaceae	*Holoptelea integrifolia* (Roxb.) Planch.						*	+	+

Flowering season: I-summer, II-monsoon, III-autumn, IV-late autumn, V-winter, VI-spring, *-indicate flowering; resources: high (+++), medium (++), low (+).

The stingless bees are generalists and polylactic foragers who visit diverse plant species to collect nectar and pollen grains (Table 9.2; Figure 9.3-9.5). However, they vigorously visited only a few plant species. Some important nectar sources are *Acacia auriculiformis*, *Borassus flabellifer*, *Brassica juncea*, *Bridelia retusa*, *Delonix regia*, *Eucalyptus tereticornis*, *Lannea coromandelica*, *Peltophorum pterocarpum*, *Syzygium cumini*, *Tectona grandis*, and *Trema orientale*. In contrast, vital pollen sources are *Brassica juncea*, *Eucalyptus tereticornis*, *Borassus flabellifer*, *Acacia auriculiformis*, *Coriandrum sativum*, *Cocos nucifera*, *Phoenix sylvestris*, *Trema orientale*. Most visited plants belong to Asteraceae, Cucurbitaceae, Euphorbiaceae, Fabaceae, Lamiaceae, Myrtaceae, Rubiaceae, and Rutaceae.

The diversity of pollen sources for the stingless bees is slightly more significant than that of nectar sources. However, they collect both the rewards (nectar + pollen) from most of the visited plants (about 69.29% of the bee-visited plants provide both nectar and pollen grains) (Figure 9.6). This is an effective foraging strategy to minimise the cost of the foraging. By collecting both nectar and pollen during their foraging trips, bees ensure a balanced diet for the colony, supporting immediate energy needs and the growth and maintenance of the hive. This dual collection also aids in pollination, benefiting the plants they visit by facilitating cross-pollination, which is vital for plant reproduction and biodiversity.

Figure 9.3. Stingless bees visiting plants. (A) *Alangium salviifolium*, (B) *Allium cepa*, (C) *Barleria cristata*, (D) *Boerhavia diffusa*, (E) *Brassica juncea*, (F) *Cajanus cajan*, (G) *Capparis zeylanica*, (H) *Capsicum annuum*, (I) *Cleome viscosa*, (J) *Combretum indicum*, (K) *Corchorus capsularis*, and (L) *Coriandrum sativum*. Scale bars = 10 mm.

Figure 9.4. Stingless bees visiting plants. (A) *Chrozophora rottleri*, (B) *Cucumis sativus*, (C) *Foeniculum vulgare*, (D) *Ixora arborea*, (E) *Jatropha gossipifolium*, (F) *Lannea coromandelica*, (G) *Lindernia crustacea*, (H) *Litchi chinensis*, (I) *Luffa acutangula*, (J) *Mangifera indica*, (K) *Momordica charantia*, and (L) *Nelumbo nucifera*. Scale bars = 10 mm.

Figure 9.5. Stingless bees visiting plants. (A) *Ocimum tenuiflorum*, (B) *Polianthes tuberosa*, (C) *Portulaca oleracea*, (D) *Psidium guajava*, (E) *Punica granatum*, (F) *Raphanus sativus*, (G) *Rosa chinensis*, (H) *Solanum melongena*, (I) *Solanum sisymbriifolium*, (J) *Turnera ulmifolia*, (K) *Trianthema portulacastrum*, and (L) *Zea mays*. Scale bars = 10 mm.

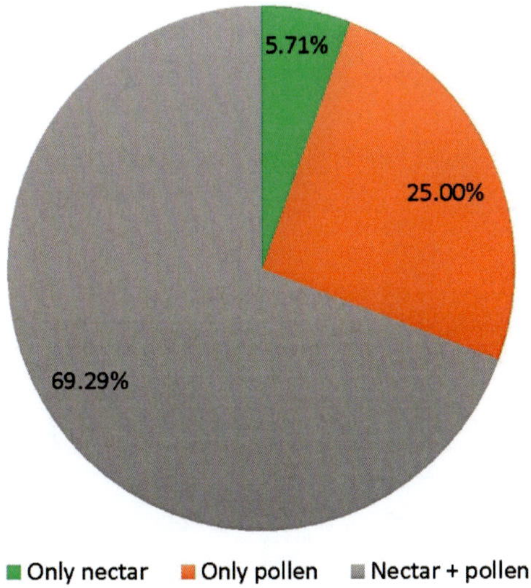

Figure 9.6. Bee-visited plants provide only nectar and pollen, as well as both nectar and pollen, to the stingless bee species.

Season-wise, the number of stingless bee-visited plants varies—more bee species visit during summer. In comparison, fewer plants are visited during autumn and late autumn (Figure 9.7). The foraging spectrum of the stingless bee species comprises wild, ornamental, crops, and horticultural plants. In semi-natural areas, the stingless bees vigorously utilise both crop plants (e.g., *Brassica juncea*, *Coriandrum sativum*, and *Momordica charantia*) and non-crop plants (viz. *Borassus flabellifer*, *Delonix regia*, *Eucalyptus tereticornis*, *Lannea coromandelica*, *Peltophorum pterocarpum*, *Tectona grandis*, and *Trema orientale*). However, the bee species mainly depend on the non-crop plants within human-altered urban areas. Progressive urbanisation with increasing population density in human-altered regions leads to a gradual decline in natural forest and agricultural fields. So, they also utilise planted trees (e.g., *Delonix regia*, *Eucalyptus tereticornis*, and *Peltophorum pterocarpum*) within the urban areas.

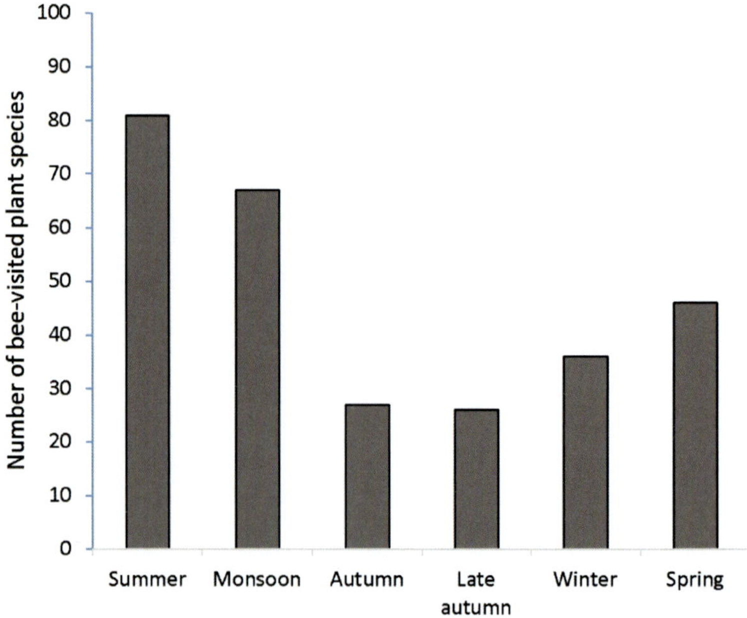

Figure 9.7. Season-wise number of bee-visited plant species.

Characteristics of nectar and pollen-providing plants

The stingless bee visits wild, ornamental, crop and fruit plants. Though most bee-visited plants are wild-type (Figure 9.8A), they visit many crop and fruit plants. Therefore, they may be essential in pollinating crops and enhancing agricultural production.

The stingless bee-visited plants are diverse regarding plant habits. Most bee-visited plants are trees (42.14%), followed by herbs, shrubs, and climbers (Figure 9.8 B). Stingless bees prefer tree blooms to those of herbs and shrubs for various reasons. Trees typically produce more significant, plentiful, and nectar-rich blooms, giving bees a greater reward for their foraging efforts. Furthermore, the height of tree flowers can provide a more secure and less disturbed habitat, away from ground-level predators and human activities. Trees also tend to bloom in large numbers, creating a visually appealing target that indicates a rich food supply. Furthermore, certain tree species have evolved to produce blossoms that appeal to bees in colour, form, and aroma, increasing their attraction to these crucial pollinators.

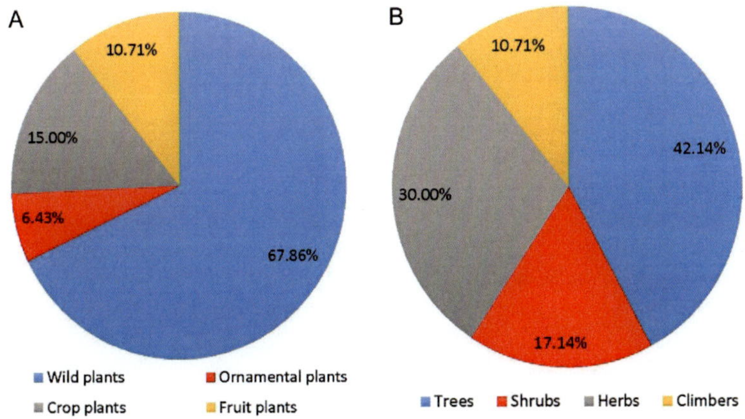

Figure 9.8. Nature of stingless bee-visited plants.

The bee-visited inflorescences are diverse, including capitulum, corymb, cyathium, cyme, fasciculate, panicle, raceme, solitary, spadix, spike, spikelet, umbel, and verticillaster (Figure 9.9 A). However, the most common are raceme (23.57%), panicle (16.43%), cyme (15.71%), umbel (15.52%), solitary (9.29%) and spike (8.57%). Most stingless bee-foraged plants were hermaphrodite (85.71%), followed by monoecious, dioecious, and polygamous. Hermaphrodite flowers can provide pollen and nectar, essential bee food sources. The availability of both floral rewards can maximise their foraging efforts, as they collect both nectar and pollen from a flower with less flight energy cost expenditure. The bee-visited flowers vary from very small (e.g., *Trema orientale*) to very large (e.g., *Luffa aegyptiaca*). Among the foraged plants, 10.71% have very small-sized flowers, 20% have small flowers, 37.86% have medium flowers, 25.71% have large flowers, and 5% have very large-sized flowers (Figure 9.9 C). Stingless bees can more easily get nectar and pollen from tiny flowers since their structures are frequently more superficial to travel, avoiding more complex floral elements that could be challenging to handle. Small flowers' morphology usually fits better with smaller bees' bodies and mouthparts. Furthermore, nectar from small flowers may be produced in sufficient amounts for small-sized bees but insufficient amounts for large-size visitors. Small bees can maximise their foraging success and minimise conflict with giant bees and other visitors by concentrating on small-

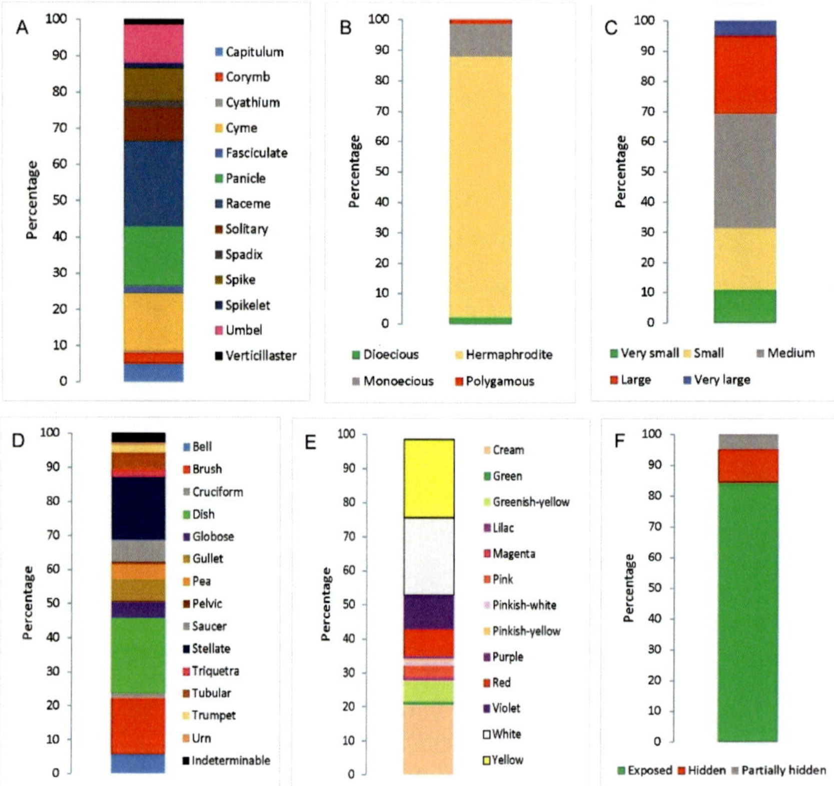

Figure 9.9. Characteristics of honeybee-visited flowers. (A) inflorescence types, (B) sexual systems, (C) flower size, (D) flower shape, (E) flower colour, and (F) pollen presentation.

sized flowers. However, large flowers can also collect resources due to exposed pollen and shallow nectaries. The bee-visited flower shapes vary, including bell, brush, cruciform, dish, globose, gullet, pea-shaped, saucer, stellate, triquetra, trumpet, tubular and urn. The predominant flower shapes were dish (22.14%), stellate (18.57%), and brush (16.43%). Dish, stellate and brush-shaped flowers are popular destinations for bees because they offer convenient access to nectar and pollen. Because of their open, flat form, dish-shaped flowers make it easy for bees to land and gather nectar without having to maneuver around intricate floral elements. Brush-shaped flowers are rewarding to foragers because they provide a wealth of pollen and nectar in a small area. They are frequently made up of several stamens or sometimes many small florets crowded together. Furthermore, their visual appeal makes these forms very apparent and alluring to bees. The colours of the flowers (as humans interpret them) visited by the stingless bees are versatile, including cream, green, greenish-purple, greenish-yellow, lilac, magenta, pink, pinkish-white, red, violet, white, and yellow. However, the most prominent bee-visited flower colour are yellow, white and cream. Bees' vision is particularly sensitive to blue and ultraviolet light, but they can also see yellow well. Yellow flowers, with their bright and vivid colour, show against the green backdrop of greenery, making them easier for bees to find. On the other hand, white flowers reflect a large quantity of light, making them visible in various lighting circumstances, including low light. Furthermore, many yellow and white flowers have evolved to feature ultraviolet patterns invisible to humans but highly appealing to bees, directing them to nectar and pollen. The pollen presentation (considering the position of anthers) of the bee-visited flowers is exposed, hidden or partially hidden (Figure 9.9F). Most bee-visited flowers have exposed anthers. Stingless bees can easily collect pollen from the exposed anthers of the flowers. The combination of visibility and extra visual signals guarantees bees can quickly locate and forage on these flowers.

CHAPTER 10

MANAGEMENT OF STINGLESS BEES

Stingless bee management strategies of the past

Stingless bees are important in social, economic, and religious factors in many parts of the world, where people have established various methods for managing and raising stingless bees. This practice has existed since prehistoric times throughout tropical America, from Mexico to Brazil (Kent 1984; Cortopassi-Laurino et al. 2006). Indigenous communities in Mexico and Latin America continue to use age-old traditional knowledge and management strategies for stingless bees. Mexico has four places where Indigenous people raise stingless bees (Dixon 1987). Traditional stingless beekeeping is related to around 18 tribes and nine subtribes in Nagaland, India (Singh 2016). People apply significant management tactics and practices in these places to harvest bee products and create new colonies. Meliponiculture has evolved over generations in Indian tribal communities with the sole intention of harvesting the products produced by stingless bees, which are an important representation of nutritional and therapeutic dietary components.

However, these traditional practices linked with meliponiculture are now only found in a few isolated places in the Mexican tropics and eastern India. The original extraction methods in the Balsas region were a form of guarded activity that systematically targeted certain nests and was considered a specialised trade, but eventually resulted in the loss of the nests (Reyes-González et al. 2016).

Early stingless bee management featured advantages and disadvantages, reflecting the balance of traditional approaches and obstacles. Additionally, these activities were environmentally sustainable, as they used low-impact approaches to protect natural habitats and increase biodiversity. The application of indigenous knowledge enabled the proper management of local bee species and vital pollinators for native flora and crops. This led to ecosystem stability and increased agricultural yield. Furthermore, early stingless bee management resulted in high-quality honey and bee products with distinct medicinal and nutritional characteristics, generating important resources for local communities. However, there were some significant drawbacks. Limited scientific understanding and resources often resulted in inefficient methods and lower yields than current techniques. Due to the reliance on traditional methods, these activities were highly localised and difficult to scale. Furthermore, there was a lack of established techniques and probable overharvesting of endangered bee populations and their habitats. Modern stingless bee conservation initiatives continue to face challenges in balancing preserving traditional knowledge with implementing new management practices.

Technical advancements in management

Since immemorial, meliponiculture has been performed only by tribal people, and scholars first hesitated to conduct honeybee studies. Despite ignorance and gaps, the practice of meliponiculture continues to this day. However, there has been a resurgence of interest in this particular activity, providing impetus for strategies to rescue traditional meliponiculture and make it more efficient through scientific studies with the goal of meliponini conservation and alternative productive projects (Reyes-González et al. 2016). This recent recovery is closely related to the increased demand for natural, organic, and homoeopathic items such as honey, pollen, propolis, and bee wax. New procedures and production processes, primarily through academic institutions, have been created over time and are known as 'integrative meliponiculture'. It is a modern technique that considers not only the production components but also the conservation of stingless bees, the integration of families interested in engaging in this activity, and the implications for community benefit (Chauhan and Singh 2019). Integrative meliponiculture aims to harvest items like honey, pollen, propolis, and cerumen from a specific nest while encouraging ecological advantages such as pollination in greenhouses and open-field crops. Several researchers manage stingless bee colonies within a hive (generally a wooden box) that can easily open to collect bee products from the nest and utilise the colony for crop pollination (Bisui et al. 2019; Layek et al. 2021, 2022). Integrative meliponiculture reduces the destruction of wild nests while promoting their maintenance, rescue, and multiplication. This action also restricts traffic and introduces non-native species into areas where stingless bees are naturally distributed. Introducing non-native bees negatively influences the foraging behaviour of local native pollinators and, thus, their survival (Layek et al. 2021).

The development of new hives for different stingless bee species (e.g., Heard 1988; Greco et al. 2010; Punith 2014) increases the rate of successful establishment of bee colonies with minimum damage to the nest during the collection of hive products. The design of the stingless bee hive also considers its portability for crop pollination purposes. For Indian stingless bees, the standard OATH hive is not suitable (author's observation). For the management of Indian stingless bees, beekeepers and researchers use different types of hives, such as single boxes with one chamber, two horizontally placed boxes, three horizontally placed boxes, or two vertically placed boxes.

Modern stingless bee management techniques have many advantages, but they also have certain drawbacks. Among the primary benefits is applying scientific knowledge and technology, which improves colony productivity and health. Improved hive designs, pest and disease management measures, and improved harvesting procedures result in higher honey and other bee product yields, allowing for more effective satisfying of commercial demands. These strategies also help with monitoring and data collection, enabling more informed decision-making and adaptive management practices. Furthermore, modern management approaches frequently incorporate breeding programs to improve desired qualities such as disease resistance and productivity, resulting in strong and resilient bee populations. Contemporary technology can also help conserve stingless bees by increasing awareness and funding large-scale conservation projects. The current stingless bee management practices may have some drawbacks. The transition to contemporary management may occasionally result in the loss of old knowledge and methods, which are essential for both the environment and culture. Additionally, there is a chance that selective breeding could homogenise bee populations, which could lower genetic diversity and the bees' capacity to adapt to shifting environmental conditions. Stingless bee management, which is becoming more commercialised and intensive, may result in the overuse of stingless bee populations, which may lead to the threat of some smaller native bees.

Different types of stingless bee hives

1. Single chamber hive

The hive consists of only one chambered box (Figure 11.1). The external hive dimensions are 220 mm × 220 mm × 220 mm. A nest entrance (about 8–10 mm in diameter) is there in the front of the colony, approximately in the middle position. The top side of the hive is openable for different purposes.

Advantages of using single chamber hive:

i. Single-chambered hives make it easy to establish new stingless bee colonies.
ii. Workers quickly and easily sealed the colony, leaving the nest entrance.
iii. The risk from predators and parasites is less.
iv. Workers need to spend less energy to maintain the colony's internal environment.
v. Easy to carry the hive to fields for crop pollination purposes.

Disadvantages of using single chamber hive:

i. The internal space of the hive is limited, and colony growth is restricted to a certain limit.
ii. It is difficult to extract hive products (e.g., honey) as brood cells, honey pots, and pollen pots are situated very closely.
iii. Difficult to colony division from a single-chambered hive.

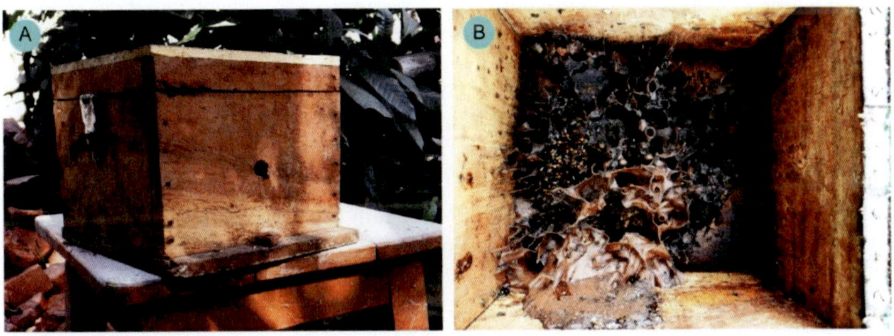

Figure 11.1. One box with a single-chambered hive. (A) External view, and (B) internal single chamber.

2. Hive with vertically separated chambers

Here the hive is one box that is separated into two chambers vertically placing a septum (Figure 11.2). The upper half portion of the septum is perforated, and the diameter of the pores is about 3-5 mm. The workers can easily move into these two chambers through the pores of the septum, but the movement of the queen is restricted. A typical hive's dimensions are 120 mm long, 150 mm in breadth, and 120 mm in height. The septum can be placed at the middle position, equally divided into two chambers, or at a place that divides into a larger chamber (2/3 portion) at the front and a smaller chamber (1/3 portion) at the back. The front chamber has a nest entrance about 8-10 mm in diameter at the middle position of the front side. The front chamber is utilised for brood rearing and also for storing food materials. The back chamber is used to store food, mainly honey.

Figure 11.2. One box with the vertically separated two-chambered hive. (A) External view, and (B) internal single chamber.

Advantages of using one box with vertically separated chambers:

i. The hive's internal space is large and has significant colony growth options.
ii. The honey chamber is isolated from the mixed brood chamber, allowing beekeepers to safely harvest hive products without injuring the brood cells.
iii. The risk from predators and parasites is less than the two or three-boxes hive.
iv. It is easier to carry the hive to fields for crop pollination purposes than the two or three-boxes hives.

Disadvantages of using one box with vertically separated chambers:

i. Workers require more time to close the nest's openings than in a single-chambered box.
ii. Workers must devote far more energy to maintaining the colony's internal environment than in a single-chambered box hive.
iii. Multiplying colonies from a one-box with a two-chambered hive is difficult.

3. Hive with horizontally separated chambers:

3.1. The hive consists of three boxes:

It looks like an OATH hive with a smaller size (Figure 3). Each segment's length, breadth, and height are 150 mm, 150 mm, and 90 mm, respectively. The bottom box has a circular entrance about 8-10 mm diameter. The chambers of the bottom and middle boxes are utilised by the stingless bees for brood rearing. At the lower side of the middle box, two split bars help separate the bottom and middle boxes. On the upper side of the middle box there is a brood excluder (a thin plate of 120 mm × 120 mm). The uppermost one is the honey box, stingless bees utilise it to store honey.

Advantages of using a three-boxes hive:

i. The internal space of the hive is high, and colony growth may reach a higher level.
ii. The honey box is separated from the brood box, so beekeepers can easily extract hive products without damaging the brood cells.
iii. The colony division is easy for a three-box hive.

Disadvantages of using a three-boxes hive:

i. A Small-sized colony of Indian stingless bees is not easily established within a three-box hive.
ii. Workers need much time to seal the openings of the nest
iii. The risk from predators and parasites is high.
iv. Workers need to spend a lot of energy maintaining the colony's internal environment.
v. Three-boxes hives are difficult to carry to fields for crop pollination purposes.

Figure 11.3. Thee-boxes hive. (A) External view showing three boxes, (B) bottom box showing the nest entrance, (C) middle box showing split bars, (D) brood excluder on the middle box, and (E) upper box or honey box.

3.2. The hive consists of two boxes:

Here, two boxes are placed horizontally, one above the other. Each segment's length, breadth, and height are 150 mm, 150 mm, and 90 mm, respectively. The bottom box has a circular entrance about 8-10 mm in diameter. The stingless bees utilise the chambers of the bottom and middle boxes for brood rearing. A brood excluder (120 mm × 120 mm) is placed between the lower and upper boxes. The chamber of the lower box is connected to the chamber of the upper box, but restricted to the entry of the queen into the upper box. So, the upper box is used for honey storage by the stingless bees.

Advantages of using a two-boxes hive:

i. The hive's internal space is large and has a significant chance of colony growth.
ii. The honey box is isolated from the brood box, allowing beekeepers to safely harvest hive products without injuring the brood cells.
iii. The risk from predators and parasites is less than the three-boxes hive.

Disadvantages of using a two-boxes hive:

i. Workers need more time to close the nest's openings than they would in a single-chamber box.
ii. Workers must expend significant energy to maintain the colony's internal environment.
iii. It is difficult to divide colonies from a two-box hive.

Multiplication of stingless bee colonies

Beekeepers can acquire new colonies by collecting nests from wild environments, trapping fresh swarms, purchasing them from other beekeepers, or multiplying their existing colonies. Colony multiplication, which involves splitting one colony into two or more colonies, is regarded as the most sustainable method of obtaining new colonies because it has low economic costs and does not need the removal of wild bees from their natural habitat (Cartopassi-Laureno et al. 2006). Colony multiplication is an integral part of stingless bee management because a large number of colonies will ensure the production of more hive products (e.g., honey, pollen, propolis, cerumen, etc.). However, colony multiplication requires precise knowledge of the best time to divide and the ability to recognise and carefully collect the brood discs at the appropriate developmental stage. Another fact is that the newly multiplied colonies are usually weak and vulnerable to pest infestations.

Queen rearing

Generally, a stingless bee colony has a queen and several drones (the number changes occasionally) that actively participate in reproduction and colony growth. They are highly eusocial and evolved a special adaptation that enable the colony to overcome queenlessness. Stingless bees breed and groom new queens throughout the year, depending on resource availability and the colony's demand. They can rear queen by two different modes- (i) rearing of larger gynes in specially built royal cells (i.e., queen cells), and (ii) rearing of gynes of similar size to that of workers in worker cells.

Factors influencing queen-rearing

Bionomic constraints of queen rearing refer to the ecological and biological factors that limit or influence the production and development of queens in social insect colonies, such as bees, ants, and wasps. These constraints can impact the success and efficiency of queen rearing, affecting the overall fitness and survival of the colony. Here are some key bionomic constraints:

1. Resource availability

The availability of resources plays a pivotal role in constraining the rearing of queens within social insect colonies, such as bees, ants, and termites. Sufficient nutrition is essential for the development of high-quality queens, as they necessitate a diet rich in proteins and fats to support their larger size and enhanced reproductive capabilities. A lack of access to these vital dietary components can reduce the number or quality of queens, which in turn impacts their survival and reproductive success. Additionally, the energy and labour required to nurture queens significantly burden the worker population, redirecting resources from other critical colony functions, including foraging, defence, and brood care. Consequently, during times of resource scarcity, colonies may be forced to prioritise immediate survival and maintenance over queen production, thereby influencing their reproductive strategies and long-term sustainability.

2.. Colony health and strength

The health and vitality of a colony are essential factors that affect the rearing of queens in social insects, including bees, ants, and wasps. A thriving and efficient colony possesses sufficient resources and a substantial workforce, enabling it to invest significantly in the production of queens. The assessment of a colony's health often hinges on the absence of diseases, parasites, and environmental stressors that could deplete vital nutrients and energy. A robust colony can allocate more workers to gather high-quality food, ensuring that the larvae intended to become queens receive optimal nutrition for their development. Additionally, a large and healthy workforce is crucial for maintaining the brood's environment, such as regulating temperature and humidity, which are vital for successful queen maturation. In contrast, colonies facing stress from illness, inadequate nutrition, or environmental challenges may find it difficult to produce queens, as resources are redirected towards survival rather than reproduction. This situation can lead to the emergence of fewer or inferior queens, thereby constraining the colony's reproductive capacity and growth. Moreover, a colony's genetic diversity and the presence of a strong queen can enhance resistance to diseases and improve resilience, thereby facilitating effective queen-rearing. In summary, the relationship between colony health, resource availability, and environmental factors significantly influences the colony's ability to generate new queens, directly affecting its reproductive success and longevity.

3. Genetic factors

Hereditary factors significantly influence queen rearing in social insect colonies, affecting both the quantity and quality of queens produced. Genetic diversity within a colony enhances its overall health and resilience, supporting successful queen-rearing. This diversity can improve disease resistance and adaptive capabilities, ensuring a stable environment for queen development. Additionally, the genetic makeup of a colony determines specific traits associated with queen development, such as the selection of certain larvae for queen rearing and the efficiency of workers in providing care. Furthermore, genetic traits can influence queen larvae's nutritional requirements and growth rates. Certain genetic lines may efficiently convert resources into reproductive success, leading to more viable and robust queens. Conversely, a lack of genetic diversity, often resulting from inbreeding, can lead to poor colony health and diminished queen quality, as deleterious alleles may become more prevalent. Therefore, genetic components play a crucial role in shaping the processes and outcomes of queen rearing, directly impacting the reproductive success and developmental fitness of social insect colonies.

4. Environmental factors

Temperature, humidity, and the availability of resources such as nectar and pollen are vital factors influencing the successful development of queens. Optimal temperatures, generally between 32–36°C, are critical for ensuring healthy larval growth and the timely emergence of a strong queen. Increased humidity levels facilitate the formation of queen cells by reducing the risk of dehydration. Furthermore, the availability of nectar and pollen directly affects the nutritional health of the colony, thereby impacting the quality and viability of the developing queen. Seasonal and climatic variations can also influence queen breeding by altering the accessibility of these essential resources and the overall health of the colony. Additionally, environmental stressors, including pesticides and diseases, complicate the queen-rearing process by introducing further challenges that may impede development and reduce the chances of successful queen replacement.

The habitat's quality significantly influences the queen-rearing process within honeybee colonies, affecting resource availability and the colony's overall health. High-quality habitats are recognised for providing diverse floral resources, ensuring a steady supply of nectar and pollen crucial for nourishing both larvae and the adult bees caring for them. Furthermore, favourable habitat conditions help to reduce exposure to various environmental stressors, such as pesticides, pollutants, and diseases, which can negatively impact bee health and the queen's developmental process. Additionally, diverse habitats create different microclimates that can protect colonies from extreme weather, thereby enhancing the resilience and success of queen-rearing. Conversely, low-quality habitats, marked by limited resources and elevated stress levels, may lead to undernourished larvae, weakened colonies, and increased susceptibility to diseases, all of which impede the production of healthy and viable queens.

5. Colony dynamics and social structure

During the growth phase, a colony emphasises expanding its population and enhancing its resources. Worker bees invest considerable effort in nurturing the brood and foraging for food. At this stage, the breeding of queens is typically limited, as the primary objective is to grow and sustain the colony. As the colony matures and enters the reproductive phase, known as swarming, the emphasis shifts to the production of new queens to establish new colonies. This phase is characterised by increased construction of queen cells and the care of multiple queens. The effectiveness of queen rearing is influenced by the dynamics within the colony, including the presence of a robust and efficient queen, the ratio of worker bees to brood, and the overall health and nutrition of the colony. A strong and stable colony with ample resources is more likely to produce high-quality queens. Conversely, a colony that experiences stress or imbalance may struggle with queen rearing, reducing the number of weaker queens. Consequently, the age of the colony and its internal dynamics play crucial roles in determining the timing, quality, and success of queen rearing.

The interactions between workers and queens are crucial in queen rearing among stingless bees, although the dynamics differ somewhat from those observed in honeybee colonies. In stingless bee colonies, workers are integral to every queen-rearing phase, including selecting larvae and nurturing their development. The pheromones released by the queen are vital for maintaining social equilibrium and for preventing workers from prematurely initiating the rearing of new queens. As the queen ages or produces a diminished quantity of pheromones, workers become increasingly motivated to start the queen-rearing process. Specific larvae are selected and provided with enhanced nutrition, typically a superior diet, to facilitate their growth into potential queens. The communication between the workers and the queen, along with the queen's pheromonal signals that establish her authority and regulate the colony, significantly influences the timing, quality, and overall success of queen rearing. Furthermore, workers may engage in activities such as constructing and protecting queen cells to ensure that the developing queens receive adequate care and can mature successfully. Thus, it is essential to maintain a delicate balance in the interactions between workers and the queen to ensure the effective development of new queens within stingless bee colonies.

6. Competition and predation

The competition among bees of the same species and different species significantly influences the production of queens in stingless bee colonies. Within a hive, bees of the same species often vie for resources, a situation that can arise from the hive's structural limitations and the availability of food and space. An overcrowded hive may lead to conflicts among worker bees over these essential resources, resulting in diminished food and space for the queen's eggs. Such competition can ultimately produce a lower quantity or quality of queens. In instances where multiple queens coexist in a hive, the rivalry among them can escalate, sometimes culminating in confrontations that may lead to the demise of weaker queens, thereby ensuring that only the most robust queen remains to govern the hive. Additionally, interactions with bees or other pollinators from different species competing for the same floral resources can further deplete the availability of vital nutrients such as nectar and pollen. This scarcity can place additional stress on the hive, hindering the nourishment of the queen's eggs and adversely affecting their development and survival. Furthermore, conflicts with other species can inflict direct harm on the hive, complicating the process of queen production. Consequently, the competition among both same-species and different-species bees is essential for the successful and effective production of queens in stingless bee colonies, ultimately impacting the overall health and viability of the hive.

The rearing of queens within stingless bee colonies encounters considerable obstacles due to predation and parasitism, which jeopardise the colony's stability and impede the successful emergence of new queens. Various predators, such as ants, birds, and other insects, directly threaten the hive by attacking and consuming adult bees, their offspring, or the developing queen larvae. This predatory pressure can provoke heightened defensive behaviours among worker bees, diverting their attention and resources from nurturing the brood and producing new queens. As a result, the overall quality and quantity of queens may decline due to insufficient care and nutrition. Additionally, parasitism, caused by organisms like mites, fungi, or parasitoid wasps, further complicates the queen-rearing process. Infestations by these parasites can compromise the hive's health, diminishing its ability to gather resources and adequately care for the brood. Parasites that specifically target developing queen larvae or pupae can induce developmental problems or deformities, leading to the emergence of non-viable queens. Furthermore, the stress associated with parasitism may compel the hive to initiate emergency queen production under less-than-ideal conditions, further diminishing the quality of the new queens. Thus, predation and parasitism present significant environmental and biological challenges that adversely influence the success of queen production in stingless bee colonies, ultimately affecting their resilience and survival.

7. Human impacts

The application of pesticides significantly influences the rearing of queens in stingless bees by introducing detrimental substances into their environment, which can adversely affect the health and development of the colony. Pesticides, particularly neonicotinoids and other systemic chemicals, can contaminate the nectar and pollen collected by bees and return them to their hive. Exposure to these tainted food sources can compromise the health of worker bees, hindering their ability to forage, nurture the brood, and produce the royal jelly vital for queen development. Even minimal pesticide exposure can induce stress in bees, resulting in disorientation, challenges in learning and memory, and a compromised immune response. This stress can diminish worker productivity and the overall vitality of the hive, complicating the production of robust queens. Furthermore, pesticides can directly impact the bees destined to become queens, potentially causing developmental issues or mortality. Such disruptions can lead to fewer queens or weaker or non-viable queens, hazarding the colony's ability to replace its queen and maintain population stability. Consequently, the use of pesticides poses a significant threat to the survival and resilience of stingless bee colonies due to its adverse effects on queen rearing.

The destruction and fragmentation of habitats significantly affect the rearing of queens in stingless bees by disrupting the availability of vital resources and the stability of their colony environment. The loss of natural habitats results in a decline in the diversity and abundance of flowering plants, thereby restricting the bees' access to nectar and pollen, essential for nourishing the colony and facilitating the development of queen larvae. Fragmented habitats lead to isolated resource patches, compelling bees to forage over longer distances and expend additional energy, which can result in nutritional stress and diminished care for the developing queens. Moreover, habitat fragmentation often leads to smaller, isolated colonies that are more susceptible to environmental changes and less capable of sustaining genetic diversity, a crucial factor for producing strong queens. These smaller colonies may face challenges in effectively rearing new queens due to the limited number of workers and resources available. Additionally, fragmented habitats heighten the risk of exposure to predators, parasites, and pesticides, which can further retard the queen-rearing efforts. The habitat loss and fragmentation create adverse conditions for stingless bees, impeding their capacity to rear healthy, viable queens and posing a threat to the long-term survival of their colonies.

Bionomic constraints for queen-rearing

- When a colony loses the queen's influence, it usually experiences significant socio-behavioural disarray.
- Mass-provisioning of brood cells necessitates the simultaneous deposit of all nutrients required to support the ontogeny of immature stages before the queen's oviposition.
- Brood cells are sealed immediately after the queen oviposits.

Artificial colony multiplication

Artificial colony multiplication for stingless bees increases their population and ensures long-term pollination services. These bees play an important role in pollinating many tropical crops and wild plants, helping to conserve biodiversity and provide food security. Natural colony multiplication rates are frequently modest, but artificial approaches can speed up colony expansion, helping to mitigate the negative effects of habitat loss, climate change, and other environmental pressures. By increasing the number of colonies, we can boost agricultural output, maintain ecological balance, and protect the genetic diversity of these important pollinators.

Various techniques, such as the splitting method (Pradeep et al. 2022) and the eduction method (Vijoykumar et al. 2013), were employed to assess their effectiveness in multiplying *Tetragonula iridipennis* colonies. The success rates across the different methods varied significantly, ranging from 0.00 to 100 percent. The lowest success rate was recorded during the separation of brood that lacked queen cells or gynes, while the highest success rate was achieved using the parallel method that included the presence of gynes.

Splitting method

For division, robust colonies with a high density of bees, a well-established brood area, and abundant supplies of pollen and honey were selected. The boxes were thoroughly cleaned, and resin was applied around the entrance to attract worker bees while deterring natural predators from entering the hive. Colony division in stingless bees was executed using two distinct methods. The first method involved dividing strong colonies that contained one or two queen cells. It was essential to transfer the queen cells to the queenless portion of the colony, while the other segment retained the existing queen. The second method involved dividing strong colonies that lacked naturally occurring queen cells. This approach stimulated the development of emergency queen cells in the queenless section of the division. The effectiveness of these two division methods and the performance of the resulting daughter colonies were evaluated against that of control colonies, which multiplied through natural swarming without manual intervention. The divided colonies were fed a honey solution mixed in a ratio of 75:25 with water at weekly intervals to promote rapid development. Food supplements enhance the quick development of the divided sister colonies.

Advantages of the splitting method

- **Rapid colony expansion:**

Splitting enables the rapid establishment of new colonies, offering a quicker alternative to eduction. The immediate division results in a more efficient means of boosting colony numbers.

- **Flexibility in timing:**

The decision to split a colony rests with the beekeeper, contingent upon the colony's strength. This discretion enables beekeepers to determine the most suitable timing for division, considering seasonal and environmental factors.

- **Control over colony formation:**

Beekeepers have greater control over the formation of the new colony. They can ensure that both the original and new colonies have the necessary resources to thrive, such as brood, honey, and pollen.

- **Effective use of strong colonies:**

Strong, healthy colonies can be split to produce new colonies without significantly weakening the original hive. This method maximises the productivity and reproductive potential of robust colonies.

- **Enhanced genetic diversity:**

By splitting colonies, beekeepers can encourage the production of new queens, promoting genetic diversity within the bee population. This can lead to stronger, more resilient colonies.

- **Immediate resource allocation:**

The new colony is created with a predetermined amount of brood, honey, and pollen. This ensures that the new colony starts with sufficient resources, reducing the risk of starvation or failure.

- **Improved colony management:**

Splitting allows beekeepers to manage colony size and prevent overcrowding, which can lead to swarming. It helps maintain a balanced bee population and reduces competition for resources within the hive.

- **Propagation of desirable traits:**

Beekeepers can selectively split colonies that exhibit desirable traits, such as high productivity, disease resistance, or gentle behaviour. This helps propagate these traits within the bee population.

Disadvantages of the splitting method

- **Colony stress:**

Splitting can cause significant stress to both the original and new colonies. The abrupt division of resources and workforce can disrupt colony dynamics and lead to temporary instability.

- **Risk of queen loss:**

The queen is at risk of being lost during the splitting process. If she is accidentally damaged or killed, it can threaten the colony's survival.

- **Resource imbalance:**

Ensuring that both the original and new colonies have adequate resources (brood, honey, pollen) can be challenging. An imbalance can lead to weakened colonies that struggle to survive and grow.

- **Dependency on colony strength:**

Only strong and healthy colonies can be split successfully. Weak or compromised colonies may not survive the process, limiting the opportunities for expansion.

- **Potential for aggression:**

Handling bees during the splitting process can provoke aggressive behaviour, leading to stings and increased risk for the beekeeper, especially if protective measures are not adequately taken.

- **Need for queen rearing:**

The new colony will need to raise a new queen or have one introduced. This requires knowledge of queen-rearing techniques and may not always be successful, leading to queen less colonies.

- **Weather dependency:**

The success of splitting can be influenced by weather conditions. Poor weather during the splitting process can hinder the bees' ability to forage and build up resources, impacting the success of the new colony.

- **Disease and pest spread:**

If not managed carefully, splitting can inadvertently spread diseases or pests between colonies. Proper hygiene and monitoring are essential to prevent the transfer of pathogens.

Eduction method

The educational approach is another way to divide and propagate stingless bee colonies. The new beehives were created from a pre-existing nest in a tree or wall that cannot be cut down or destroyed. This method involves removing a new hive from the original colony at the fixed location. The eduction method can also be used to propagate artificial hives, but it requires more time and patience than the splitting method. In the case of a preexisting nest within a tree or wall, it cannot be removed or destroyed. The original nest is still in place and the process of duplicating it can be done annually. Robust and vigorous colonies were chosen based on the foraging abilities. The opening of the colony entrance was delicately taken out using a pinprick technique and replaced with a black plastic tube (10 mm in diameter) that was situated in the same place as the original entrance. This caused all incoming and outgoing foragers to pass through the newly installed plastic opening. The entrance to the original nest was securely attached to the end of the plastic tube to allow the bees to easily recognize the familiar scent of their old entrance. The pheromone from the fresh hive is spread inside and outside the entrance, serving as a beacon at the entryway and reducing disturbances among the bees. The bees were effortlessly moved to the entrance of their nest and any disturbance was kept to a minimum. After one week, the original entrance tube was taken off the tip and replaced with a new wooden box hive measuring 200 mm × 100 mm × 100 mm. It was made of 15 mm thick wooden planks with two 20 mm entrance holes in the centre of both the front and back panels. The new hive was placed in the same spot as the parent colony. Ensure that the box hive is adequately supported when placing it near the feral colony. All the bees had to leave their original nest and move into the annexe hive through the new plastic entrance. The openings and cracks are tightly sealed, making it impossible for the bees to get out of the tube without passing through the attached hive. At first, bees from three colonies utilised the additional space within the annexe hive to store waste and began creating unique cerumen layers near the entrance. The diluted honey (1:1) was placed in the mud lamp feeder and a group of pollen pots were added to the annexure hive to attract the bees from the original colony to the annexed hive. Artificial honey was first brought into the main nest and stored inside, which led to increased building activity with the addition of food pots and brood cells. These circumstances frequently lead to a significant increase in the size of the brood nest. These circumstances continued to contribute to the colony's increasing strength at a relatively fast pace, leading to a rise in the temperature of the brood nest. The higher number of worker bees and increased nest temperature led to a group of worker bees occupying the vacant area of the annexe. Additionally, the bees began building their food stores. Later on, the bees stored their extra food in the additional hive. A fully developed queen cell containing worker brood from a healthy and robust colony was transferred to the annexe hive. The nurse bees in the annexure hive began building brood cells for the newly mated queen bee. A thriving bee colony can establish a daughter colony within three to five months in an area with ample food resources and a favourable climate. However, in areas with unsuitable environmental conditions and limited food resources, it may take a year or more for the colony to reproduce, or it may fail to do so altogether. Carefully separate the annexure hive from the parent colony and seal all holes of the annexure. The enclosed daughter colony was relocated to a new spot after nightfall. The efficacy of the education program depends on the diversity of workers in different age groups and the presence of larval and pupal cells in the colony. The educational approach took place when there was ample food supply, particularly pollen. This could influence bees to construct brood cells for the queen to lay eggs, which may be crucial for success. The available space in the new hive will only be used when there is limited space for further nest expansion within the wild colony.

Advantages of the eduction method

- **Minimal disturbance:**

Eduction involves creating a new colony by attracting bees from an existing hive into a new empty hive. This process is less disruptive compared to traditional splitting methods, which can cause significant stress to the bees.

- **Natural colony growth:**

Eduction allows for more natural growth of the new colony. Bees gradually move to the new hive, bringing along brood, honey, and pollen, ensuring a balanced development of the new colony.

- **Preservation of the original colony:**

The original colony remains largely intact and can continue to thrive and produce. This is beneficial for beekeepers who want to expand their colonies without significantly weakening the existing ones.

- **Reduced risk of queen loss:**

In eduction, the queen stays in the original hive, and the new colony will raise its own queen. This reduces the risk of losing a queen during the transfer process, a common issue in traditional splitting methods.

- **Enhanced hive health:**

By allowing bees to move at their own pace, the eduction method promotes better hive health. Bees can maintain their natural behaviours and hive structure, which contributes to overall colony well-being.

- **Better adaptation:**

The new colony established through eduction is better adapted to its new environment, as bees have moved and settled voluntarily, ensuring that the new hive site meets their requirements.

Disadvantages of the eduction method

- **Time-consuming:**

The process of eduction can be slow, as it relies on the bees gradually moving into the new hive. This can take several weeks to months, which may not be ideal for beekeepers looking for quicker colony expansion.

- **Dependency on environmental conditions:**

Eduction success is highly dependent on favourable environmental conditions. Poor weather, lack of forage, or other stressors can slow down or even halt the process, making it less reliable than other methods.

- **Initial setup requirements:**

Properly setting up the eduction process can be tricky. It requires positioning the new hive close enough to the existing colony to encourage movement but not so close that it causes confusion or interference between the two hives.

- **Risk of incomplete colony formation:**

There is a risk that the new colony may not form completely or may lack sufficient resources, such as brood or honey stores, to sustain itself, especially if the migration is incomplete.

- **Monitoring and intervention:**

Although less interventionist than some methods, eduction still requires regular monitoring to ensure that bees are moving into the new hive and that the new colony is developing properly. This can be time-consuming and may require corrective actions.

- **Variable success rates:**

The success of eduction can be inconsistent. Some colonies may adapt well and thrive, while others may struggle to establish themselves. This variability can make eduction a less predictable method compared to more direct techniques.

- **Initial hive strength requirements:**

The original colony must be strong and healthy to support the eduction process. Weak or stressed colonies may not be able to spare enough bees and resources to establish a new colony successfully.

CHAPTER 11

STINGLESS BEE NEST PRODUCTS

Different types of stingless bee hive products

Stingless bee foragers collect nectars and pollen grains from flowers and other sugary substances. They produce honey from nectar or other sugary substances. Stingless bees store honey and pollen within their hive for the growth and development of the colony (Figure 11.1).

Stingless bees collect plant resin and latex and add wax, saliva, and other substances to produce propolis and cerumen. Although propolis and cerumen are sometimes used interchangeably in the literature on stingless bees, these are two distinct nesting materials. Cerumen has a greater wax content and is soft and less sticky than propolis. Propolis is found mainly in sealing areas of the nest, while cerumen is found in food pots, involucrum, and other supporting strands (Figure 11.1). When soil is blended with resin and/or latex, the product is known as 'geopropolis' or 'batumen'. Some publications use the word geopropolis interchangeably with stingless bee propolis, but this is inaccurate because most Meliponini bees do not add clay to the resin and/or latex (Celemli 2013). Indian stingless bees (also known as dammar bees) occasionally employ clay in their nests and form geopropolis (Figure 11.2). Sometimes, stingless bees can store latex/resin (a comparatively lighter colour than propolis and cerumen) inside the hive for future use (Layek et al., 2023). These bee products can be collected from the wild nest or managed hive by opening the nest.

Figure 11.1. Hive products. (A) Honey pots, (B) honey, (C) pollen pots, (D) pollen loads, (E–F) propolis, and (G) cerumen.

Figure 11.2. A wild stingless bee nest shows geopropolis surrounding the nest entrance.

Honey

What is honey, and how is it formed?

Honey is a naturally sweet substance made by honeybees or some stingless bees from flower nectar, secretions of living plant components, or plant-sucking insect excretions. Honeybees and some stingless bees produce honey. Stingless bee honey is known by several names, including Meliponine honey, stingless bee honey, pot honey, and Kelulut honey (in Malaysia).

Honey is produced through a fascinating and intricate process involving worker bees' cooperative efforts. It all starts with forager bees collecting nectar from flowers (and other sugary liquids) using their long, tube-like mouths. The nectar is collected and stored in a specialised stomach known as the "honey stomach," which is separate from the bee's digestive system. The bee's body enzymes (secreted from their cephalic glands) break down the complex carbohydrates in the nectar into simpler sugars like glucose and fructose as it is transported back to the colony. When a forager bee returns to the hive, it regurgitates the nectar into the mouth of a house bee, transmitting it from bee to bee through a process known as "trophallaxis." During this exchange, the nectar is mixed with enzymes, which break down more complex carbohydrates and aid in the evaporation of water. This repetitive procedure gradually thickens the nectar, resulting in honey. The house bees then put the partially digested nectar into the honey pots, where they fan it with their wings to accelerate the evaporation of water and reduce the moisture content. This reduction in water content inhibits fermentation, preserving and stabilising the honey for long-term storage. When the honey has reached the correct viscosity, the bees seal the honey pots with cerumen to preserve it from moisture and contamination. The outcome is honey, a thick, sweet substance that is an essential food resource for the colony, particularly during times of scarcity when foraging cannot be conducted.

Physicochemical properties of stingless bee's honey

Stingless bee honey is characterised by its unique flavour and aroma, exhibiting a more liquid consistency and a gradual tendency to crystallise. Nevertheless, prolonged storage in honeypots may lead to crystallisation (Figure 11.3), which is contingent upon the honey's composition. The lack of comprehensive understanding regarding its composition has resulted in a deficiency of identity and quality standards for stingless bee honey, which diverges from the established criteria set forth by the international honey standards of the Codex Alimentarius Commission (2001), specifically tailored for honey produced by *Apis* species (Silva et al. 2013; Chuttong et al. 2016). Stingless bee honey is contained within cerumen pots, and consequently, its quality is affected by the infiltration of phytochemicals from the cerumen (Temaru et al. 2007; Massaro et al. 2011). Within these cerumen pots, the nectar collected undergoes three distinct transformation processes before it is converted into honey.

Figure 11.3. Honey crystallisation within the honey pots.

Similar to *Apis* honey, the honey produced by stingless bees comprises water, carbohydrates, organic acids, phenolic compounds (including phenolic acids and flavonoids), proteins, amino acids, enzymes, vitamins, and minerals (Habib et al. 2014). The following section outlines the physicochemical properties of stingless bee honey.

- **Moisture content:**

  The moisture content of honey depends on bee species, nectar sources, and weather conditions. Generally, stingless bee honey has a greater water content than the honey of *Apis* spp. (Bijlsma et al. 2006; Souza et al. 2006). Moisture content ranges from 22-34%. The moisture content has exceeded the limits specified by international standard regulation, which is less than 20% for Apis honey. The moisture content is important in honey for determining its shelf-life (Lage et al. 2012; Shahnawaz et al. 2013). High moisture content may lead to the fermentation of honey and the loss of flavour, which will cause the honey to lose its quality. However, honey deterioration also depends on the chemical composition of honey. High moisture content may slow down the crystallisation of sugars present in honey.

- **Ash content:**

Ash content in honey may represent minerals in honey, although in a very small amount present. This could be the indicator of the botanical and geographical origin of the honey. Ash content of stingless bee honey ranges from 0.02-0.07% (w/w). The value is lower than the values reported for other stingless bee species (Chuttong et al. 2016: 0.53 ± 0.63 g/100 h honey).

- **pH:**

The pH value of stingless bee honey is higher than that of Apis honey, ranging from 3.20 to 4.52. A similar pH value was reported for the honey of different stingless bee species (Souza et al. 2006; Chuttong et al. 2016). A low pH value may inhibit microorganisms' growth and thus prevent honey's deterioration (Do Nascimento et al. 2015).

- **Acidity:**

The acidity of the stingless bee honey ranges from 44.2-195.8 meq/kg. International honey standard (2002) recommended a maximum of 50 meq/kg of *Apis* honey acidity. Most stingless bee honey significantly exceeds the limit value. A high amount of acidity is also reported in the honey of other stingless bee species (Lage et al. 2012; Chuttong et al. 2016; Ya'akob et al. 2019). This could be the outcome of the fermentation process.

- **Electrical conductivity:**

The electrical conductivity of stingless bee honey ranges from 0.15 to 2.80 ms/cm. The electrical conductivity of honey is associated with the ash content (mineral content) and acidity, revealing the presence of ions, organic acids, and proteins (Silva et al. 2016). Thus, this parameter has been used as a honey quality indicator, assisting in identifying and distinguishing floral honey and honeydew (Lazarevic′ et al. 2012; Karabagias et al. 2014). The maximum electrical conductivity of honey derived from *Apis* spp. was determined by the International Honey Commission in 2002 to be less than 0.8 mS/cm for blossom honey and greater than 0.8 mS/cm for honeydew honey.

- **Total soluble solids (TSS):**

Soluble solids in honey are compounds that dissolve in water, such as sugars, acids, proteins, phenols, salts, and organic molecules (Colucci et al. 2016). Honey's total soluble solids can be assessed using a Brix° unit refractometer. Indian stingless bee honey has total soluble solids ranging from 65.90 to 80.40 Brix°, similar to values reported for other stingless bee species (Sousa et al. 2016; Ya'akob et al. 2019). Brix°, which measures sugar levels in honey, can be used to detect adulteration (Guerrini et al., 2009). The soluble solids indicate sugar, organic acid, and mineral content, but they are also directly related to sugar and water levels in the samples.

- **Carbohydrates:**

The total carbohydrate content in stingless bee honey ranges from 63.50% to 75.70%. Almost similar carbohydrate content is found in other stingless bee honey (e.g., Suntiparapop et al. 2012: 66 ± 3.8 g/100 g honey). Generally, dietary fibre is not found in honey. The dominant sugar is maltose (about 37 g/100 g honey), fructose (about 17 g/100 g honey), and glucose (about 14 g/100 g honey). Other sugars, like turanose, trehalose, and raffinose, are present in small quantities. The composition of the sugars depends mainly on the nectar sources and bee species. consequently, there is a broad range of variation in the fructose-to-glucose ratio (about 0.4 to 2.5). Honey that is high in fructose typically stays liquid for several months or even years. In contrast, the crystallisation of glucose-rich honey can begin shortly after harvesting, as glucose is significantly less soluble in water than fructose.

- **5-Hydroximetilfurfural (HMF):**

The 5-Hydroxymethylfurfural (HMF) is a recognisable parameter to determine the freshness of honey. HMF content is highly vary among honey samples, from 0.25 mg/kg to 8.7 mg/kg. The increasing HMF content may be due to several factors, such as storage conditions, pH, time and temperature of heating, and floral sources (de Almeida-Muradian et al., 2013). HMF often occurs due to the acid-catalyzed dehydration of hexose sugars (fructose and glucose) or from Maillard's reaction (Al-Diab and Bushra Jarkas 2015).

- **Protein content:**

Most honey contains a low amount of protein compared to other food sources. Even stingless bee honey has a low protein content, like *Apis* honey. Protein content may range from 0.17% to 0.92%. The proteins present in honey are primarily enzymes (e.g., glucose oxidase, diastase, and invertase).

- **Enzyme activity:**

The main enzymes in honey include glucose oxidase, diastase, and invertase. Glucose oxidase facilitates the transformation of glucose into gluconic acid and hydrogen peroxide, the latter contributing to the antimicrobial characteristics of honey. Diastase is responsible for the breakdown of starch into simpler sugars, while invertase acts to hydrolyse sucrose into glucose and fructose, which are the predominant sugars present in honey. The efficacy of these enzymes can fluctuate due to various factors, including the floral origin of the nectar, the methods employed in honey processing, and the conditions under which it is stored. Over time, the activity of these enzymes may diminish, particularly when honey is subjected to heat or light, which can adversely affect these delicate proteins. This enzymatic activity is crucial in determining the flavour, texture, health benefits, and quality indicators associated with honey production. For the Indian stingless bee honey, the diastase is about 31.20 DN (diastase number), and the invertase is about 136.15 IN (invertase number) (Divya et al. 2018). Chuttong et al. (2016) reported that the diastase activity in stingless bee honey varied from 0.05 to 4.9 °Gothe, with a mean value of 1.5 ± 1.6 °Gothe. In contrast, Vit (2013) presented a broader range of diastase activity for honey from South American stingless bees, spanned from 2.4 to 21 °Gothe.

- **Minerals:**

The mineral profile of stingless bee honey depends on its botanical origin and the stingless bee species (Souza et al. 2006). The most concentrated mineral element is potassium (about 576.6 ± 177.6 ppm, Fuenmayor et al. 2012), followed by calcium, copper, iron, magnesium, sodium, and zinc.

- **Phenolic compounds:**

  The profile of phenolic compounds of stingless bee honey shows wide variation. The phenolic composition of honey depends on floral and geographical origin, as well as the time of collection. Major phenolic compounds in honey are gallic acid, caffeic acid, p-coumaric acid, kaemferol, naringin, taxifolin, 2-hydroxycinnamic acid, salicylic acid, 2,4-dihydroxybenzoic, ferulic, elagenic, myricetin, and chrysin. The total phenol content in Indian stingless bee honey ranges from 516.95 mg catechol/kg to 1318.20 mg catechol/kg, with an average of about 827.42 mg catechol/kg. The flavonoid content ranges from 92.85 mg catechin/kg to 374.22 mg catechin/kg, with a mean value of 196.60 mg catechin/kg (Divya et al. 2018).

Antioxidant activity of stingless bee's honey

The honey's antioxidant activity assessment employs various analytical techniques, focusing on distinct antioxidant constituents. Among these methods is the DPPH (2,2-diphenyl-1-picrylhydrazyl) assay, which determines honey's capacity to neutralise free radicals by observing the reduction in absorbance as DPPH is diminished. Another prevalent technique is the ABTS (2,2'-azino-bis(3-ethylbenzothiazoline-6-sulfonic acid)) assay, which generates a blue-green ABTS radical cation and measures the decline in absorbance as the antioxidants present in honey counteract the radical. Furthermore, the FRAP (Ferric Reducing Antioxidant Power) assay evaluates honey's reducing ability by tracking the conversion of ferric ions (Fe^{3+}) to ferrous ions (Fe^{2+}), resulting in a coloured complex. The ORAC (Oxygen Radical Absorbance Capacity) assay gauges the capacity of antioxidants to absorb and neutralise peroxyl radicals, thereby indicating honey's potential to mitigate oxidative stress. These methodologies are frequently supplemented by spectrophotometric techniques to quantify the total phenolic and flavonoid content, which are closely associated with antioxidant activity.

Generally, stingless bee honey contains more flavonoids and polyphenols than the honey produced by *Apis* spp. The major factors that contribute to the antioxidant content in honey are phenolics, organic acids, peptides, enzymes, Maillard reaction products, and possibly other minor components (Berreta et al., 2005). Though honey composition varies from zone to zone, the average antioxidant content is 374.84 uM Fe (II) per 100 g of honey. The ascorbic acid content ranges between 1.60 to 5.20 mg/kg, with an average value of 2.70 mg/kg.

Antimicrobial activity of stingless bee's honey

The honey produced by stingless bees can inhibit the proliferation of both gram-positive and gram-negative bacteria (Pimentel et al. 2013). A distinctive characteristic of this honey is its storage in honey pots constructed from cerumen, which exhibits notable antimicrobial properties (Choudhari et al. 2012). The antibacterial properties of honey can be categorised into peroxide and non-peroxide components (Molan 2009). While non-peroxide activity is rarely found in honey from *Apis* species, Temaru et al. (2007) have identified significant non-peroxide activity in various samples of stingless bee honey. Furthermore, Indian stingless bee honey demonstrates considerable antibacterial activity against several bacterial strains, including *Bacillus subtilis*, *Escherichia coli*, *Pseudomonas aeruginosa*, and *Salmonella typhi* (Figure 11.4; authors' observation).

Modern uses of stingless bee honey

Culinary uses

Stingless bee honey is increasingly recognised in the culinary sector for its unique flavour and adaptability. In contrast to traditional honey, stingless bee honey typically presents a tangy, mildly sour taste accompanied by a rich and intricate flavour profile, rendering it a sought-after component in gourmet cuisine. It serves as a natural sweetener in various beverages, desserts, and sauces, where its distinctive flavour contributes depth and character to the dishes. Culinary professionals frequently utilise it in salad dressings, marinades, and glazes to elevate the taste of both sweet and savoury preparations. Furthermore, its lower glycemic index relative to standard honey makes it a favoured option among health-conscious individuals, who regard it as a more nutritious substitute for sugar. Its capacity to harmonise with other ingredients while delivering a unique flavour makes stingless bee honey an esteemed addition to contemporary culinary innovations.

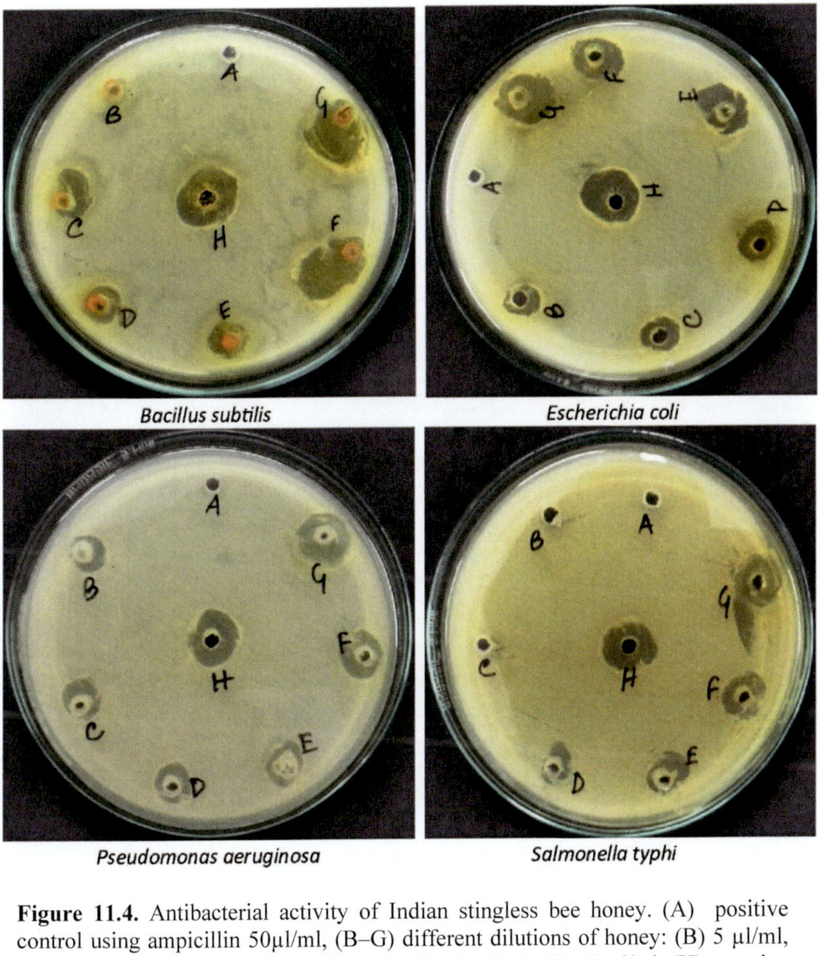

Figure 11.4. Antibacterial activity of Indian stingless bee honey. (A) positive control using ampicillin 50µl/ml, (B–G) different dilutions of honey: (B) 5 µl/ml, (C) 10 µl/ml, (D) 20 µl/ml, (E) 30 µl/ml, (F) 40 µl/ml, (G) 50 µl/ml, (H) negative control using streptomycin 50µl/ml.

Nutritional uses

Stingless bee honey is gaining recognition for its impressive nutritional properties, serving as a substantial source of natural sugars, vitamins, minerals, and antioxidants. In contrast to traditional honey, it typically boasts elevated levels of phenolic compounds, which deliver significant antioxidant benefits, aiding in the protection against oxidative stress and promoting overall well-being. Its lower glycemic index renders it a preferable sweetener for individuals managing blood sugar levels, including those with diabetes or those aiming to decrease their sugar consumption. Furthermore, stingless bee honey is rich in trace elements such as zinc, iron, and magnesium, enhancing its nutritional profile and supporting immune health, energy metabolism, and overall metabolic function. It also acts as a natural energy enhancer, offering a rapid source of easily digestible carbohydrates, making it particularly beneficial for athletes and physically active individuals. In summary, the comprehensive nutritional attributes of stingless bee honey make it a significant addition to a well-rounded diet.

Medicinal applications

Stingless bee honey has been esteemed for its medicinal benefits for centuries, and contemporary studies are further investigating its therapeutic applications. This honey is particularly noted for its robust antimicrobial characteristics, rendering it an effective natural solution for treating wounds, burns, and skin infections. Its capacity to enhance tissue regeneration and inhibit bacterial proliferation expedites wound healing. Moreover, stingless bee honey exhibits significant anti-inflammatory and antioxidant effects, which are advantageous in the management of chronic inflammatory diseases such as arthritis while also supporting overall health by counteracting detrimental free radicals. It is also utilised in addressing gastrointestinal issues, including gastritis and ulcers, which aids in soothing and safeguarding the digestive system. Furthermore, stingless bee honey is gaining recognition for its potential in diabetes management due to its lower glycemic index and ability to help regulate blood sugar levels. As a natural remedy, stingless bee honey is a versatile and powerful asset in enhancing health and well-being.

Skin and hair care

Stingless bee honey is a highly beneficial component in skin and hair care, attributed to its exceptional moisturising, healing, and nourishing qualities. Functioning as a natural humectant, it effectively attracts moisture to the skin, making it an ideal ingredient in face masks, lotions, and creams for hydration and rejuvenation. Its inherent antimicrobial and anti-inflammatory characteristics assist in addressing acne and various skin infections, thereby fostering clearer and healthier skin while alleviating redness and irritation. In hair care, stingless bee honey is incorporated into conditioners and hair masks to improve shine, softness, and manageability. Additionally, it aids in alleviating dryness and dandruff by hydrating the scalp and enhancing overall hair vitality. Whether integrated into skincare or hair care regimens, stingless bee honey provides a natural and efficient means to enhance beauty and support well-being.

Cultural and religious significance

Stingless bee honey possesses considerable cultural and religious significance across numerous societies globally. It is often viewed as a representation of purity, abundance, and prosperity, frequently incorporated into various rituals and ceremonies. For example, in particular indigenous communities in Southeast Asia and Latin America, stingless bee honey plays a vital role in traditional healing methods and spiritual practices, invoking blessings and promoting well-being. Its importance is also evident in religious observances, where it is presented as a sacred offering or utilised in holy foods, highlighting its esteemed status as a pure and beneficial substance. Furthermore, in traditional medicinal systems such as Ayurveda and indigenous practices in South America, stingless bee honey is revered for its alleged health benefits. It is employed to address a range of health issues, thereby further integrating it into cultural and spiritual traditions. This cultural relevance emphasises the profound relationship between stingless bee honey and the communities that value its various applications and symbolic interpretations.

Bee pollen

Bee pollen is a mass of pollen packed by worker honeybees into the comb cells. Bee pollen is harvested as food for humans. Its nutritional composition comprises proteins, lipids, sugars, fibres, mineral salts, amino acids, and vitamins (Wesh and Marston 1983; Marchini et al. 2006). In addition, pollen also has high polyphenolic content, chiefly flavonoids with antioxidants (Campos et al. 2003) and antimicrobial activity (Basim et al. 2006). Bee pollen has been used for traditional medicine, supplementary nutrition, and alternative diets (Serra Bonvehi and Escola Jorda 1997; Isla et al. 2001). Bee pollen is safe for short-term use, but side effects include allergic reactions. Bee pollen is unsafe for pregnant women and should not be used during breastfeeding.

Propolis

Chemical constituents of propolis

Indian stingless bee propolis contains various chemicals such as alkenes, aliphatic acids, aromatic acids, thiophilic acids, sugars, esters, and terpenes (Table 11.1). Kothai and Jayanthi (2014) found that the total phenol content of propolis (ethanolic extract) was 150 µg/ml of Gallic acid equivalent (GAE), and the flavonoid content was 6 mg/g of Quercetin equivalent. Kasote et al. (2019) found that the ethanolic extract sample's total phenolic and flavonoid content ranged from 9.63–48.60 µg GAE/mg and 0.30–4.93 µg QE/mg. Propolis contains significant amounts of gallic acid (0.27–1.27 µg/mg) and naringin (0.24–1.16 µg/mg). P-coumaric acid and kaempferol were also detected in cerumen samples.

Table 11.1. Chemical constituents of propolis of Indian stingless bees.

Chemical compounds	Empirical formula	Molecular weight	Source
Alpha-[p-bromophenyl]-o-amino cinnamic acid	$C_{11}H_{10}BrNO_3$	284.11	Choudhari et al. 2012
1,2-Benzenedicarboxylic acid	$C_8H_6O_4$	166.13	Choudhari et al. 2012
β-Arabinose, TMS	$C_{17}H_{42}O_5Si_4$	438.85	Choudhari et al. 2012
Caffeic acid	$C_9H_8O_4$	180.16	Kasote et al. 2019
3-Cyclohexenyl acetate	$C_8H_{12}O_2$	140.18	Choudhari et al. 2012
1,2-Dihydro-4,6-dimethyl-2-oxo, nicotinic acid	$C_{10}H_{13}NO_3$	195.22	Choudhari et al. 2012
2,5-Dihydrothiophene	C_4H_6S	86.16	Choudhari et al. 2012
3,3- Dimethylpentane	C_7H_{16}	100.21	Choudhari et al. 2012
Eremophilene	$C_{15}H_{24}$	204.36	Choudhari et al. 2012
Ethyl diazoacetate	$C_4H_6N_2O_2$	114.10	Choudhari et al. 2012
Fructose, MEOX-5TMS	$C_{22}H_{55}NO_6Si_5$	570.10	Choudhari et al. 2012

Gallic acid	$C_7H_6O_5$	170.12	Kasote et al. 2019
3-Hydroxybutanoic acid	$C_4H_8O_3$	104.11	Choudhari et al. 2012
Hydroxysteric acid phenacyl ester	$C_{24}H_{40}O_3$	376.30	Choudhari et al. 2012
Kaempferol	$C_{15}H_{10}O_6$	286.23	Kasote et al. 2019
2-Methyl, 1-penten-3-yne	C_6H_8	80.13	Choudhari et al. 2012
Naringin	$C_{27}H_{32}O_{14}$	580.54	Kasote et al. 2019
Nitromethane	CH_3NO_2	61.04	Choudhari et al. 2012
p-Coumaric acid	$C_9H_8O_3$	164.16	Kasote et al. 2019
5-Phenoxymethyl-1,3,4-thiadiazol-2-amine	$C_9H_9N_3OS$	207.25	Choudhari et al. 2012
Ribitol, 5TMS	$C_{20}H_{52}O_5Si_5$	513	Choudhari et al. 2012
(RR)-(+), 3,3,4-trimethyl-4-p-tolyl-cyclopentanol	$C_{15}H_{22}O$	218.33	Choudhari et al. 2012

Antimicrobial activities of propolis

Researchers utilise ethanolic extract (Choudhari et al. 2012; Kothai and Jayanthi 2014), methanol-water extract (Surendra et al. 2012), and chloroform or acetone extract (Brumfitt et al. 1990) to detect propolis' antibacterial activity. Propolis extract is highly effective against Gram-positive and Gram-negative bacteria and yeast. In comparison to the honeybees, stingless bee propolis exhibits stronger antibacterial activity *against specific bacteria (e.g., Bacillus cereus, Burkholderia glumae, Erwinia nigrifluens, Klebsiella* sp., *Vibrio logie*, and *Xanthomonas campestris*) (Surendra et al. 2012). Propolis (methanol-water extract) had a reduced minimum inhibitory concentration (MIC) against *Burkholderia glumae* (60%), *Bacillus cereus* (75–80%), and *Xanthomonus campestris* (75–80%). The ethanolic extract of propolis inhibits selected microbes (*Aeromonas* sp., *Escherichia coli*, *Klebsiella* sp., *Salmonella* sp., *Staphylococcus* sp., and *Vibrio parahaemolytics*) at lower concentrations (500 μm/ml), with a maximum zone of inhibition (5–12 mm) observed at 1000 μm/ml (Kothai and Jayanthi 2014). Choudhari et al. (2012) found that MICs and MBCs ranged from 1.21 to 9.75 μg/ml and 2.43 to 19.5 μg/ml. Kasote et al. (2019) demonstrated that an ethanolic extract of propolis had substantial antibacterial activity in the order *Candida albican* > *Staphylococcus aureus* = *Bacillus subtilis* > *Pseudomonas aeruginosa* > *Escherichia coli*. Cerumen components may block bacterial DNA-dependent RNA polymerase (Simuth et al. 1986), or flavonoids may be involved (Surendra et al. 2012). However, Kasote et al. (2019) discovered that MIC values of samples correlated negatively with both total phenolic and flavonoid content, implying that molecules other than phenolics and flavonoids, such as terpenes, may be responsible for propolis' antimicrobial action.

Antioxidant properties of propolis

The antioxidant activities of propolis samples can be assessed using DPPH (2,2- diphenyl-1-picrylhydrazyl) radical scavenging ability, ABTS [2,2-azino-bis(3-ethylbenzothiazoline-6-sulfonic acid)], and FRAP (ferric reducing antioxidant power) assays. DPPH is a stable free radical compound, deep violet, showing maximum absorbance at 517 nm. When DPPH is added to the cerumen extract, it gets reduced. Upon reduction of this radical by an antioxidant, the absorbance at 517 nm will be disappeared. In the ABTS assay, cerumen extracts will be allowed to react with $ABTS^{\bullet+}$ solution (produced by reacting between ABTS and potassium persulfate in a ratio of 1:1) in a dark condition. Then, the absorbance will be recorded at 734 nm using a spectrophotometer. FRAP assay is based on the reduction of Fe^{3+} TPTZ complex (a colourless complex) to Fe^{2+} tripyridyl triazine (a blue-coloured complex) formed by the action of electron-donating antioxidants. The change in absorbance at 593 nm is used to track the process.

Propolis' phenolic and flavonoid content positively correlates with its free radical scavenging action. The ethanolic extract of propolis has an antioxidant capacity of 83%. Kasote et al. (2019) found that the antioxidant activity of propolis samples (ethanolic extract) ranged from 0.05 to 0.07 mM Trolox/mg in the DPPH assay, 0.04–0.59 mM Trolox/mg in the ABTS assay, and 0.03–0.21 mM Trolox/mg in the FRAP assay.

Anticancer activity of propolis

Propolis (ethanolic extract) was investigated for anticancer activity using cytotoxic and apoptotic effects in various cancer cell lines, including

B16F1 (murine melanoma), Caco-2 (human epithelial colorectal adenocarcinoma), HT-29 (human colon adenocarcinoma), and MCF-7 (human breast cancer). The cytotoxicity of cerumen extract can be assessed using the MTT [3-(4,5-dimethylthiazol-2-yl)-2,5-diphenyl tetrazolium bromide] assay and the Trypan blue dye exclusion assay. The apoptotic effect can be assessed by Acridine Orange/Ethidium Bromide (AO/EB) staining and the TUNEL (terminal deoxynucleotidyl transferase dUTP nick end labelling) assay.

Choudhari et al. (2013) found that cerumen extract (at higher doses) dramatically lowers the viability of cancer cell types. Propolis ethanolic extracts at low concentrations (< 50 μg/mL) were noncytotoxic to cancer cell lines. They found an IC_{50} value of 250 μg/mL for the investigated cell lines. At that concentration, considerable numbers of apoptotic cells were discovered. All cell lines exhibited apoptosis-related morphology (nucleus shrinkage, cell shape changes, membrane blebbing, nuclear fragmentation, and chromatin condensation). Cells treated with ethanolic extract of propolis (250 μg/mL) showed high TUNEL positivity (>99%).

Geopropolis

The Indian stingless bees produce geopropolis by mixing wax, pollen, the mandibular secretions of bees, plant resins and latex, and the addition of soil. Geopropolis is utilised within the hive to seal cracks, define the spaces occupied by bees, and restrict excessive airflow. This substance comprises various compounds, including benzoic acids, benzophenones, phenolic substances such as terpenes, hydrolysable tannins, flavonoids, cinnamic acid, vanillin, and long-chain fatty acids like palmitic and oleic acids (Dutra et al. 2014; Santos et al. 2017). The bioactive compounds present in geopropolis are believed to contribute to its diverse biological effects, which include antimicrobial (Cunha et al. 2013), anti-inflammatory (Franchin et al. 2016), antinociceptive (Franchin et al. 2012), gastroprotective (Ribeiro-Junior et al. 2015), antioxidant (Dutra et al. 2014), antiproliferative (Cunha et al. 2016), antimutagenic (Santos et al. 2017), and cytotoxic (Bartolomeu et al. 2016) properties.

Cerumen

Little information is available on the chemical compositions and biological activity of stingless bee cerumen. The chemical composition is more similar to that of stingless bee propolis, albeit a little difference can be noticed with a higher wax content. Cerumen contain phenolic compounds, fatty acids, terpenoids, and minerals. Common phenolics are gallic acid, catechin, vanillin, kaempferol, rutin, 3,4-hydroxycinnamic acid and maleic acid. Fatty acids are capric acid, lauric acid, myristic acid, palmitic acid, stearic acid, myristoleic acid, oleic acid, linolenic acid, and linoleic acid. Among metallic minerals commons are aluminium, calcium, potassium, magnesium and iron. Non-metallic minerals are phosphorous and sulfur. Besides these, xanthonoids (e.g., alpha-mangostin) are also present in cerumen. The total phenolic content varies from 3.70 to 81.50 mg GAE/g of raw cerumen.

The methods (e.g., DPPH$^{\bullet}$ and ABTS$^{\bullet+}$ free radical scavenging assays, FRAP assay) show a significant antioxidant activity of cerumen extract. The antioxidant activity of cerumen is mainly provided by gallic acid and kaempferol (Deyerling et al. 2024). A few compounds (e.g., tomentosenol A) have antimicrobial activities. The alpha-mangostin exhibits strong anticancer activity against a broad spectrum of human cancer-derived cell lines. Its cytotoxicity was substantially more potent against malignant than normal cells (Nugitrangson et al., 2016). Some components of cerumen (e.g., tomentosenol A) have significant anti-fibrotic activity (Hamilton et al. 2022).

CHAPTER 12

STINGLESS BEES PROVIDE POLLINATION SERVICES

Pollination and pollinators

Pollination is the process by which pollen grains are transferred from a flower's male reproductive organs (anthers) to the female reproductive organs (stigma) of the same or another flower of the same species. This transfer of pollen is essential for fertilisation and the production of seeds and fruit in flowering plants.

Types of pollination (mode of pollen transfer)

1. **Self-pollination:** Self-pollination occurs when pollen from a flower's anther is transferred to the stigma of the same flower or another flower on the same plant (Figure 12.1).
2. **Cross-pollination:** Cross-pollination occurs when pollen from a flower's anther is transferred to the stigma of a flower on a different plant of the same species.

Different types of fertilisation modes

1. **Autogamy:** Autogamy, also known as self-fertilisation, occurs when pollen from the anther of a flower fertilises the ovules within the same flower.
2. **Allogamy:** Allogamy is the fertilisation that occurs when pollen from the anther of one flower fertilises the ovules of a different flower, either on the same plant or on a different plant of the same species. It may be two types-
    2.1. **Geitonogamy:** Fertilization occurs by the pollen transfer of one flower to another of the same plant.

2.2. **Xenogamy:** Fertilization occurs by the pollen transfer of one flower to another of a different plant of the same species.

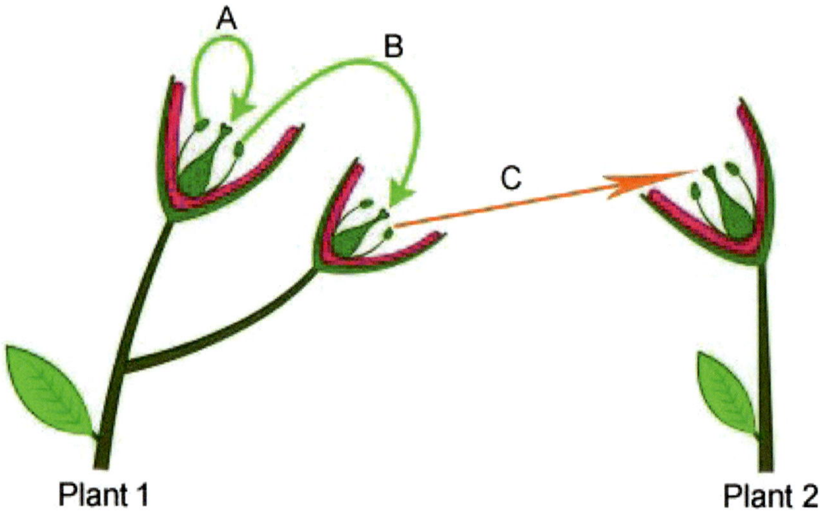

Figure 12.1. Different types of pollination. (A-B) self-pollination, (C) cross-pollination.

Pollination can occur through various agents, including abiotic (e.g., wind and water) and biotic (such as insects, birds, and mammals), or self-pollination within the same flower. Based on the pollinating agents, pollination is of the following types:

1. **Anemophily:** Anemophily refers to pollination carried out by wind. In this pollination, pollen grains are released into the air and transported by wind currents to land on the stigma of other flowers. Plants adapted to anemophily often produce large quantities of lightweight pollen.
2. **Hydrophily:** Hydrophily is a relatively rare type of pollination that occurs in aquatic plants. Water currents carry pollen to the stigmas of other flowers that are submerged.

3. **Zoophily:** Zoophily is pollination facilitated by animals. This includes insects (entomophily) and birds, bats, and sometimes other mammals. These animals visit flowers in search of nectar, and as

they feed, they pick up pollen on their bodies, which they then transfer to other flowers as they move between them.

a. **Entomophily:** This type of pollination involves the transfer of pollen by insects. Bees, butterflies, moths, flies, beetles, and insects visit flowers to feed on nectar or pollen, inadvertently transferring pollen between flowers as they move.
b. **Chiropterophily:** Chiropterophily is pollination facilitated by bats. Bats often pollinate night-blooming flowers, typically white or pale-coloured, emitting strong fragrances to attract bats in the darkness. Bats feed on nectar and inadvertently transfer pollen between flowers.
c. **Ornithophily:** Ornithophily specifically refers to pollination by birds. Birds like hummingbirds have long beaks and tongues adapted to extract nectar from tubular flowers. As they feed, they pick up and transfer pollen between flowers.

Features make stingless bees as good pollinators

1. **Eusocial colony:**

Stingless bees are eusocial, living in a colony with many foragers. The number of foragers and the number of visits received to a flower are positively correlated.

2. **Hairy body:**

Stingless bees, like other bees, have hairy bodies. The hairs help them hold pollen on their body surface and brush pollens on flower stigmas.

3. **Generalise flower visitors:**

Stingless bees are polylectic and generalise visitors of many wild and cultivated plants (Layek and Karmakar 2018). They provide pollination services to many plant species.

4. **Floral constancy:**

Stingless bees visit one type of flower (a single plant species) on a single bout or trip. Even daily, an individual forager collects the same resources throughout bouts. However, depending on resource availability, the resource-collecting tasks may show flexibility (Layek et al. 2021).

5. **Legitimate type of visits:**

Stingless bees mostly visit flowers as legitimate visits, i.e., touch reproductive parts of the flowers. Thereby, they efficiently pollinate the flowers.

6. **Flower visitation rate and handling time:**

Like other bees, stingless bees have a high flower visitation rate compared to dipteran, hemipteran, and lepidopteran members. They also spend significant time on flowers during resource collection. Unlike beetles, bugs, butterflies, and flies, they do not like to spend time on other plant parts than flowers. Therefore, they can pollinate many flowers within a short time.

Why do we need supplementary bee pollination?

Multiple factors, including habitat loss, pesticide use, climate change, diseases, invasive species, pollution, monoculture farming, and lack of awareness and conservation efforts drive the decrease in pollinator populations. Most crop plants depend on pollinators partially or obligatorily (e.g., members of Cucurbitaceae). Even fruit and seed sets of some anemophilous crops are positively influenced by insect visits. Due to pollinator decline, crops face a pollination deficit that negatively influences yield qualitatively and quantitatively. For the following major issues, the management of bees is required:

i. Intensive agriculture

Intensive agriculture demands managed pollinators due to its reliance on high-yield monoculture crops, which require concentrated and efficient pollination services within a short timeframe. These farming practices often involve vast expanses of single crops like almonds, apples, and berries that bloom simultaneously, creating a high, localised demand for pollinators that natural populations cannot meet. Additionally, these fields' uniform and large-scale nature reduce the availability of diverse forage and habitat necessary for sustaining wild pollinator populations throughout the year. Pesticides and other chemicals commonly used in intensive farming further diminish wild pollinator numbers and health. Managed pollinators, such as honeybees, can be strategically placed and moved to different fields as needed, providing a reliable and controlled pollination service that ensures crops are adequately pollinated, leading to higher yields and better-quality produce. This predictability is crucial for farmers who depend on consistent and timely pollination to maximise crop productivity and economic returns. Moreover, managed pollinators are regularly monitored and treated for diseases and pests, maintaining their health and efficiency. This level of management and control is essential to meet the specific and intense pollination needs of modern intensive agriculture, supporting the production of food at the scale required to feed a growing global population.

ii. Supplementing declining wild populations

Wild pollinators are declining due to factors primarily driven by human activities. Habitat loss and fragmentation from urbanisation, agricultural expansion, and deforestation reduce the availability of natural foraging and nesting sites for pollinators. Pesticide use, particularly neonicotinoids, harms pollinators directly and indirectly by contaminating their food sources. Climate change alters the timing of flowering plants and disrupts the synchrony between pollinators and their food resources while also affecting pollinator health and distribution. Additionally, the spread of diseases and parasites, often exacerbated by globalisation and the movement of managed bee colonies, further threatens wild pollinator populations. These combined stressors create a challenging environment for wild pollinators, leading to their decline and raising concerns about the impacts on biodiversity and ecosystem services.

Even in areas where wild pollinators are present, their numbers and activity can be unpredictable. Managed bees provide a reliable backup, ensuring crops are pollinated even if wild populations are insufficient.

iii. Economic factors

Adequate pollination by managed bees leads to higher crop yields, as more flowers are successfully pollinated, resulting in a greater number of fruits, seeds, or nuts per plant. This increase in yield directly translates to higher quantity, ensuring that farmers can produce enough to meet market demands. Additionally, well-pollinated crops often exhibit superior quality, with fruits and vegetables being more immense, uniform, and visually appealing, enhancing their market value. The economic benefits of managed bee pollination are substantial, as increased yields and improved quality lead to higher income for farmers. Moreover, reliable pollination services provided by managed bees reduce the risk of crop failures and ensure a more stable and predictable production cycle, further contributing to economic stability. The overall boost in agricultural productivity also supports the broader economy by sustaining food supply chains, creating jobs, and contributing to export revenues. Thus, managed bee pollination is critical in optimising agricultural outputs and driving economic growth in the farming sector.

iv. Flexibility in utilisation

Managed bee colonies offer the invaluable advantage of being deployable at any time and in any season, ensuring that crops receive essential pollination services regardless of the time of year. This flexibility is particularly beneficial for modern agriculture, where different crops have varied blooming periods and may require pollination at specific times to maximise yield and quality. Beekeepers can move managed bee colonies to various locations as needed, aligning with the flowering schedules of diverse crops such as almonds in early spring, apples in mid-spring, and squash in the summer. This ability to provide consistent and timely pollination supports continuous agricultural production and allows farmers to cultivate various crops throughout the year. Furthermore, managed colonies can be overwintered or maintained in controlled environments, ensuring their availability and health even during off-seasons or adverse weather conditions. This year-round applicability of managed bees enhances the resilience and productivity of agricultural systems, helping to secure food supply and support global food security.

Managed bees, particularly stingless bee colonies, are highly mobile, allowing beekeepers to transport their colonies to various crop fields for pollination. This mobility is crucial for ensuring timely and effective pollination of diverse crops that bloom at different times or regions. By moving bee colonies to where they are most needed, farmers can ensure that crops receive optimal pollination during their critical flowering periods, leading to higher yields and improved produce quality. This flexibility also helps overcome the challenges posed by the decline of wild pollinator populations, as managed bees can be deployed to areas where natural pollinators are scarce or insufficient. Additionally, the ability to transport bees helps maintain their health and productivity, as beekeepers can move colonies to areas with abundant forage and better environmental conditions. This strategic mobility of managed bees is a cornerstone of modern agricultural practices, significantly enhancing the efficiency and reliability of pollination services and thereby supporting sustainable and productive farming systems.

v. Global food security

Global food security is intricately linked to managed bee pollination because many of the world's essential crops rely on pollinators for fruit, seed, and nut production. Approximately 75% of global food crops benefit from animal pollination, with bees being the most significant contributors. Managed bee pollination ensures that crops like almonds, apples, blueberries, and numerous vegetables receive the consistent and effective pollination they need to produce high yields and quality produce. This is particularly crucial in intensive agricultural systems where large-scale monocultures cannot rely on dwindling wild pollinator populations to meet their pollination needs. The ability to strategically place and move managed bee colonies allows for precise pollination during critical blooming periods, optimising crop production and ensuring food availability. As the global population continues to grow, the demand for food increases, necessitating reliable and efficient agricultural practices to sustain food supply. Managed bee pollination directly supports this by enhancing crop yields, diversifying the types of crops that can be successfully grown, and contributing to the stability and resilience of food production systems. Furthermore, by securing the pollination of essential crops, managed bees help mitigate the risk of food shortages and contribute to economic stability in agricultural sectors. In regions where food security is already a challenge, the role of managed pollinators becomes even more critical, providing a sustainable means to boost farm productivity and support local and global food systems.

Indian stingless bees enhance crop yield: some case studies

Bell pepper (*Capsicum annum* L.)

Stingless bees' foraging activity on capsicum begins at 7.00 h and continues up to 18.00 h. Although bell pepper is largely self-pollinated, bee pollination enhances the fruit set and quality (Table 12.1). The fruit set in stingless bee-pollinated capsicum remains higher (89.57%) than the pollinator-exclusion treatment (82.43%) (Rakshitha 2024). The stingless bee pollination system increases fruits' length, girth, and weight. Stingless bee pollination also improves the fruit quality, with more total soluble solids (TSS) content (5.21%) and ascorbic acid content (93.41 mg/100 g) than the pollinator exclusion system.

Table 12.1. Influence of stingless bee pollination on bell pepper's fruit set and fruit quality.

Parameter	Pollinator-exclusion system	Stingless bee pollination system
Fruit set (%)	82.43	89.57
Fruit length (mm)	6.39	7.43
Fruit girth (mm)	18.77	22.11
Fruit weight (g/fruit)	82.93	98.31
Yield (g/plant)	0.45	0.69
TSS content (%)	4.43	5.21
Ascorbic acid (mg/100g)	89.70	93.41
Moisture (%)	79.60	81.87

Bitter gourd (*Momordica charantia* L.)

The pollination and fruit set of bitter gourd is obligatory and depends on insect pollinators, which are predicted by the pollinator exclusion system where fruit sets do not occur (Table 12.2). Pollination service to the crop is performed by several native pollinators (honey bees, stingless bees, solitary bees, butterflies, and wasps). The crop suffered from pollination deficit (coefficient of pollination deficit, D = 0.20). Therefore, the percentage of the fruit set of bitter gourd was significantly lower in the open pollination system (75 ± 12.69) compared to the maximum potential, i.e., the fruit set obtained in the hand-pollination system (94 ± 8.43). The use of managed stingless bees in bitter gourd pollination increases the percentage of fruit set in comparison to the open pollination system. Fruit quality also significantly improved by managed stingless bee pollination compared to the open-pollination treatment. Length, girth, and weight of fruits were higher in the managed bee-pollinated system than in the open-pollination system (Table 12.2).

Table 12.2. Effect of different pollination treatments on field-grown bitter gourd (Bisui et al. 2020).

Parameter	Pollination treatments			
	Pollinator-exclusion	Open-pollination	Hand-pollination	Managed stingless bee pollination
Fruit set (%)	0	75 ± 12.69	94 ± 8.43	86 ± 11.74
Fruit quality is measured 10 days after the flower opens				
Length (mm)	-	108.40 ± 7.17	122.9 ± 8.31	115.30 ± 8.84
Girth (mm)	-	80.30 ± 2.41	96.7 ± 5.81	92.60 ± 5.32
Weight (g)	-	25.90 ± 2.83	35.45 ± 3.84	30.91 ± 5.39

Cucumber (*Cucumis sativus* L.)

The stingless bee pollination on cucumber increases the fruit set more than the open-pollination system (Sen et al. 2023). The length, girth, and weight of fruits also increase by managed stingless bee pollination (Table 12.3). Cucumber yield remains higher in the stingless bee pollination system (13.15 ± 6.68 tonnes per hectare) and increases by about 58.62% compared to the open-pollination system (i.e., non-manipulated system).

Table 12.3. Fruit set and yield of cucumber in different pollination treatments.

Parameter	Open-pollination	Stingless bee pollination
Fruit set (%)	66.03 ± 12.69	79.06 ± 8.60
Fruit length (cm)	14.65 ± 0.94	18.08 ± 1.23
Fruit girth (cm)	8.13 ± 0.47	10.08 ± 0.59
Fruit weight (g)	145.77 ± 7.73	202.62 ± 12.98
Yield (tonne/hector area)	8.29 ± 1.60	13.15 ± 6.68

Fennel (*Foeniculum vulgare* Mill.)

Many insects (ants, bugs, butterflies, flies, honeybees, solitary bees, stingless bees, and wasps) visit fennel flowers (Layek et al. 2022). Most visitors touched the reproductive parts of the fennel flowers and were regarded as legitimate visitors. During the morning, floral visitors, especially *Tetragonula iridipennis*, collected pollen grains and visited mostly freshly opened flowers. In the afternoon (2.00–6.00 pm), they act mainly as nectar foragers and visit both the current-day flowers and one-day-old flowers (with senesced petals and stamens).

Considering the percentage of fruit set among the three pollination treatments (viz. pollinator-exclusion, open-pollination, and manual cross-pollination treatments), the highest fruit set obtained in the manual cross-pollinated treatment, followed by open-pollination and the least amount of fruit set in the pollinator-exclusion treatment (Table 12.4). In the open system, the coefficient of pollination deficit D is 0.19.

Table 12.4. Impact of managed stingless bee colonies on floral visitors and crop yields of fennel (Layek et al. 2022).

Impacts	Pollinator-exclusion	Open-pollination (without managed bee)	Managed stingless bee colonies
On the abundance of floral visitors			
Number of total visitors/m² area/5 min	-	23.30 ± 9.80	26.58 ± 10.35
Number of *Tetragonula iridipennis*/m² area/5 min	-	1.34 ± 1.33	5.24 ± 3.42
On crop yield			
The year 2020			
Fruit set (%)	$48^c \pm 15.49$	$72^b \pm 12.29$	$83^a \pm 9.49$
Number of fruits per compound-umbel	$102.85^c \pm 49.85$	$179.55^b \pm 86.57$	$202.40^a \pm 83.00$
Fruit weight (g)	$0.50^b \pm 0.03$	$0.65^a \pm 0.03$	$0.67^a \pm 0.03$
Yield per plant (g)	$10.59^c \pm 4.23$	$24.72^b \pm 6.54$	$29.38^a \pm 6.17$
Yield per hectare (kg)	$414.4^c \pm 26.55$	$877.47^b \pm 134.77$	$1061.33^a \pm 152.33$
The year 2021			
Fruit set (%)	$46^c \pm 15.78$	$69^b \pm 11.97$	$79^a \pm 11.97$
Number of fruits per compound-umbel	$98.20^c \pm 53.20$	$173.90^b \pm 86.52$	$195.10^a \pm 82.60$
Fruit weight (g)	$0.49^c \pm 0.03$	$0.63^b \pm 0.03$	$0.66^a \pm 0.03$
Yield per plant (g)	$9.68^c \pm 3.84$	$22.47^b \pm 5.81$	$26.59^a \pm 5.82$
Yield per hectare (kg)	$379.47^c \pm 13.54$	$836.8^b \pm 98.50$	$984^a \pm 105.87$

Values are given as mean ± standard deviation. Values followed by different letters differ significantly by Tukey-b post hoc test ($p < 0.05$).

The utilisation of managed stingless bee colonies increases the abundance of floral visitors on fennel. It thereby increases fruit set by about 14.89% and 72.34% higher than the open-pollination (i.e., without managed bees) and pollinator-exclusion treatments, respectively. The number of matured fruits per compound-umbel increases in the managed stingless bee pollination system compared to the open-pollination and pollinator-exclusion treatments. The fruit weight of fennel also depends on the degree of pollination services perception. The fruit weight remains higher in supplementary pollination through managed stingless bee colonies and native pollinators. The crop yield (regarding fruit weight per plant and per hectare field area) remains higher in the managed stingless bee pollination compared to the open-pollination and pollinator-exclusion systems (Table 12.4). Installing managed stingless bee colonies within the fennel fields increased yield by about 19.31% compared to the open-pollinated system (Layek et al. 2022).

Watermelon (*Citrullus lanatus* (Thunb.) Matsum. & Nakai)

Many insects (Beetles, butterflies, flies, honeybees, solitary bees, stingless bees, and wasps) visit watermelon flowers, with an abundance of *Halictus acrocephalus* and *Lasioglossum funebre* in open-field conditions (Layek et al. 2021).

Table 12.5. Fruit set and fruit quality of watermelon in different pollination systems (Layek et al. 2021).

Measure	Pollinator-exclusion	Open-pollination	Cross-pollination	Managed stingless bees
Fruit set (%, n = 20×5)	0^{d}	$64^{c} \pm 6.52$	$88^{a} \pm 4.47$	$78^{b} \pm 10.37$
Fruit quality (after 40 days of flower open)				
Length (cm, n = 30)	-	$24.48^{c} \pm 3.26$	$28.07^{a} \pm 1.99$	$26.04^{b} \pm 2.24$
Girth (cm, n = 30)	-	$63.20^{c} \pm 5.81$	$70.10^{a} \pm 3.55$	$66.37^{b} \pm 4.19$
Weight (kg, n = 30)	-	$3.01^{c} \pm 0.40$	$3.72^{a} \pm 0.26$	$3.36^{b} \pm 0.40$

Values are given as mean ± standard deviation. Means in the row followed by the same letters do not differ significantly by DMRT at 5%.

The percentage of fruit set remains the highest in the manual cross-pollination system and lowest in the open-pollination system (Table 12.5). Under open conditions (i.e., non-manipulated system), the coefficient of pollination deficit (D = 0.27) indicates a significant pollination limitation for watermelon. With the introduction of stingless bees for watermelon pollination, the fruit set increased by about 14%. Crop yield is increased by approximately 36.05% by applying managed stingless bees.

CHAPTER 13

PROSPECTS OF STINGLESS BEES

Cosmopolitan distribution

Stingless bees belonging to the tribe Meliponini within the family Apidae are predominantly found in tropical and subtropical regions worldwide. They thrive in Central and South America, Africa, Southeast Asia, and Australia. These bees exhibit diverse nesting behaviours, from tree hollows to intricate nests, and play crucial roles as pollinators in various ecosystems. Central and South America, particularly the Amazon rainforest, boast rich diversity in stingless bee species. Additionally, Africa, Southeast Asia, and Australia harbour significant populations of these bees. Despite their concentration in tropical regions, some species are also found in subtropical areas like parts of the United States. Overall, stingless bees contribute significantly to ecosystem health and biodiversity through pollination.

Easily manageable

Stingless bees are without functional stinging apparatus and are less hazardous to human beings. Stingless bee management is relatively more accessible compared to honeybees for several reasons:

- **Non-aggressive behaviour:** Stingless bees are generally less aggressive compared to honeybees. They lack a potent sting, which reduces the risk of stings to beekeepers and makes handling them less intimidating.
- **Smaller colonies:** Stingless bee colonies are typically smaller than honeybee colonies. This makes managing them more manageable, requiring less space and resources.
- **Propolis seals:** Stingless bees seal their nests with propolis, a resinous substance that helps regulate temperature and humidity within the hive. This natural seal reduces the likelihood of pests

and diseases affecting the colony, minimising the need for beekeeper interventions.

- **No need for protective gear:** Due to their non-aggressive nature and lack of a powerful sting, beekeepers often do not need protective gear when working with stingless bees. This simplifies hive inspections and other management tasks.
- **Indigenous knowledge:** In many regions where stingless bees are native, traditional knowledge and practices regarding their management exist. Indigenous communities often have established methods for keeping and harvesting stingless bee colonies, making management practices more accessible and culturally integrated.
- **Nesting preferences:** Stingless bees have specific nesting preferences, often utilising tree hollows, cavities, or man-made hives designed to mimic natural nests. Providing suitable nesting sites can encourage colony establishment and facilitate management.
- **Low maintenance requirements:** Stingless bee colonies typically require lower maintenance than honeybee colonies. They do not produce surplus honey in the same quantities as honeybees, which means less frequent harvesting and extraction activities for beekeepers.
- **Pest and disease resistance:** Stingless bees have evolved mechanisms to resist pests and diseases commonly affecting honeybees. While they can still be susceptible to certain pathogens and parasites, these issues are generally lower in stingless bee colonies.

Why need stingless bee management?

Pollinator populations are decreasing due to habitat loss, pesticide use, climate change, diseases, invasive species, monoculture agriculture, urbanisation, and light pollution. The decline in pollinators poses significant threats to ecosystems, agriculture, and human well-being. It disrupts plant reproduction, reduces crop yields, threatens food security, compromises biodiversity, and affects economic sectors reliant on pollination. Additionally, it can impact human health, cultural traditions, and ecosystem aesthetics. Traditionally, honeybee management has been popularised for many years to overcome pollinator loss and their hive products. However, exotic honeybee domestication negatively impacts local pollinators (Layek et al. 2021; Page and Williams 2023). Therefore, the management of indigenous stingless bees is necessary to address the following issues:

- **Pollination services:** Stingless bees are effective pollinators of various plants, including many native and crop species. Managed colonies can enhance pollination in agricultural settings, leading to increased crop yields and improved quality of fruits and seeds.
- **Conservation:** Many stingless bee species face threats from habitat loss, pesticide exposure, and competition with introduced bee species. Managed colonies can contribute to the conservation of these bees by providing safe nesting sites, food resources, and protection from threats.
- **Biodiversity:** Stingless bees are an integral part of ecosystems, playing roles in pollination, seed dispersal, and nutrient cycling. Managing stingless bee populations helps maintain biodiversity by supporting the reproduction of diverse plant species and sustaining ecological interactions.
- **Cultural and traditional practices:** Stingless bees are significant in many indigenous communities, where they are valued for their honey, wax, and medicinal properties. Managed colonies can help preserve traditional beekeeping practices and maintain cultural heritage.
- **Research and education:** Managed stingless bee colonies provide opportunities for research and education. Studying their behaviour, biology, and ecology can enhance our understanding of pollination dynamics, insect conservation, and ecosystem functioning. Educating the public about the importance of stingless bees can also promote their conservation and sustainable management.
- **Economic benefits:** Stingless bee products such as honey, wax, and propolis have economic value. Managed colonies can generate income for beekeepers through the sale of these products. Additionally, stingless beekeeping can create employment opportunities in rural areas and contribute to local economies.
- **Alternative pollinators:** With concerns about declines in honeybee populations due to factors like colony collapse disorder, managing stingless bees provides an alternative pollination option for farmers and growers. Diversifying pollination services helps mitigate the risks associated with relying solely on honeybees for crop pollination.

Outputs of stingless bee management

Stingless bee management yields a range of valuable outputs, including honey, wax, and propolis production. Managed colonies contribute to pollination services, enhancing crop yields and biodiversity conservation. They also preserve cultural traditions and indigenous knowledge while raising awareness about pollinator conservation. Additionally, stingless bee management fosters research opportunities, leading to advancements in beekeeping practices and conservation strategies. Overall, it generates economic, environmental, and social benefits, highlighting the significance of sustainable beekeeping practices. The outputs of stingless bee management encompass various tangible and intangible benefits, including:

- **Hive products:**
  Stingless bees produce honey, often highly prized for its unique flavours and medicinal properties. Managed colonies can yield honey harvested sustainably for consumption or sale, contributing to income generation for beekeepers. In addition to honey, stingless bees produce wax and propolis. Wax can be harvested for various purposes, including candle-making, cosmetics, and pharmaceuticals. Propolis, a resinous substance collected by bees from plant sap, has antimicrobial properties and is used in traditional medicine and health products.

- **Pollination services:**
  Managed stingless bee colonies contribute to pollination in agricultural and natural ecosystems. By visiting flowers to collect nectar and pollen, stingless bees facilitate the reproduction of many plant species, including crop plants. Improved pollination can lead to increased crop yields, improved fruit quality, and enhanced ecosystem stability.

- **Crop yield enhancement:**
  In agricultural settings, managed stingless bee colonies can lead to higher crop yields and improved fruit sets. Farmers may experience increased productivity and profitability through enhanced pollination services provided by stingless bees.

- **Biodiversity conservation:**
  Managed stingless bee colonies maintain ecosystem biodiversity through their role as pollinators. By supporting the reproduction of

diverse plant species, stingless bees help sustain food webs and ecosystem functions, ultimately contributing to the conservation of natural habitats and species diversity.

- **Cultural and traditional values:**
  Stingless bee management often involves preserving cultural traditions and indigenous knowledge of beekeeping. Managed colonies may be valued for their honey and other products and their cultural significance in local communities, where they are often integrated into traditional practices and ceremonies.

- **Research opportunities:**
  Managed stingless bee colonies provide valuable scientific research and innovation opportunities. Researchers may study various aspects of stingless bee biology, behaviour, ecology, and management techniques, leading to new insights and advancements in beekeeping practices and conservation strategies.

Limitations of stingless bee management

Stingless bee management faces several limitations, including low honey production, limited market demand, dependency on wild populations, challenging management practices, limited research and resources, climate sensitivity, risk of hybridisation, and regulatory constraints. These factors hinder the development of stingless beekeeping and pose challenges for sustainability and commercial viability. However, ongoing research, education, and collaboration efforts aim to address these limitations and improve stingless bee management practices for long-term sustainability. While stingless bee management offers numerous benefits, it also presents several limitations:

- **Low honey production:**
  Stingless bees produce relatively small quantities of honey compared to honeybees. This limits the potential for commercial honey production and may not meet the demand in specific markets.

- **Limited market demand:**
  Stingless bee honey, wax, and propolis may have niche markets and limited demand compared to honeybee products. Marketing

these products effectively can be challenging, especially in regions where stingless bee products are not well-known or valued.

- **Dependency on wild populations:**
  In many regions, stingless beekeeping relies on capturing wild colonies rather than breeding domesticated strains. This dependence on wild populations raises concerns about sustainability and conservation, as overharvesting can deplete wild bee populations and disrupt ecosystem dynamics.

- **Challenging management practices:**
  Stingless bee management techniques may be less developed than honeybee management practices. Limited knowledge and experience in stingless beekeeping can result in challenges such as colony losses, pest and disease outbreaks, and suboptimal hive conditions.

- **Limited research and resources:**
  Stingless bees receive less research attention and funding than honeybees. As a result, there may be gaps in knowledge about stingless bee biology, behaviour, and management, hindering efforts to develop effective management strategies and address challenges.

- **Climate sensitivity:**
  Some stingless bee species are sensitive to changes in temperature and humidity, which can affect colony health and productivity. Climate change and extreme weather events may exacerbate these challenges, posing risks to managed colonies.

- **Risk of hybridisation:**
  Introducing managed colonies of stingless bees into areas with native populations can pose risks of hybridisation and genetic introgression, potentially compromising the genetic integrity of local bee populations.

- **Regulatory constraints:**
  Regulation and restrictions on stingless bee management and trade vary between regions and countries. Compliance with regulatory requirements can pose administrative burdens and limit opportunities for stingless beekeepers.

Current scenario of stingless bee domestication

Early domestication of Indian stingless bees (*Tetragonula iridipennis*) was started by tribal people (Kani tribe) of Western Ghats, India, using bamboo bee hives (Kumar et al. 2012). Recently, researchers and beekeepers conducting meliponiculture with the stingless bee species in several states of India, including Kerala, Tamil Nadu, and West Bengal. Yet, a standard bee box is not fixed for the bee species. Instead, they utilise different types of artificial wooden boxes. Extraction of hive products (mainly honey) may damage the nest severely. To avoid this problem, researchers are using hives with two partially separated chambers (horizontally and vertically) to manage the stingless bee species. Besides hive products, managed stingless bee populations are utilised for crop pollination, and in most cases, supplementary bee pollination increases crop yields (Bisui et al. 2020; Layek et al. 2021,2022b; Sen et al. 2023). Due to growing available data, easy management, higher market value of bee products, and agricultural yield perspective, the management of stingless bees is progressively increasing in Indian states, including West Bengal.

REFERENCES

Al-Diab D, Jarkas B. 2015. Effect of storage and thermal treatment on the quality of some local brands of honey from Latakia markets. Journal of Entomology and Zoology Studies 3(3): 328–334.

Araújo ED, Costa M, Chaud-Netto J, Fowler HG. 2004. Body size and flight distance in stingless bees (Hymenoptera: Meliponini): inference of flight range and possible ecological implications. Brazilian Journal of Biology 64: 563–568.

Baker EW, Delfinado M. 1985. An unusual species of Neocypholaelaps (Acari: Amerosiidae) from Asia. International Journal of Acarology 14: 127–136.

Bartumeus F. 2007. Levy processes in animal movement: An evolutionary hypothesis. Fractals 15(2): 151–162.

Balaji K, Jayaraj J, Shanthi M, Vellaikumar S, Rajamanickam C, Chitra N, Suresh K. 2023. Morphometrics of Inrraspecific populations of stingless bee *Tetragonula irirdipennis* Smith from southern Tamil Nadu. Indian Journal of Entomology 86(1): 1–4.

Bartolomeu AR, Frión-Herrera Y, Da Silva LM, Romagnoli GG, De Oliveira DE, Sforcin JM. 2016. Combinatorial effects of geopropolis produced by *Melipona fasciculata* Smith with anticancer drugs against human laryngeal epidermoid carcinoma (HEp-2) cells. Biomedicine & Pharmacotherapy 81: 48–55.

Basim E, Basim H, Özcan M. 2006. Antibacterial activities of Turkish pollen and propolis extracts against plant bacterial pathogens. Journal of Food Engineering 77(4): 992–996.

Beretta G, Granata P, Ferrero M, Orioli M. 2005. Standardization of antioxidant properties of honey by a combination of spectrophotometric/ fluorimetric assays and chemometrics. Analytical Chemistry 533: 185–191.

Bijlsma L, De Bruijn LIM, Martens EP, Sommeijer MJ. 2006. Water content of stingless bee honeys (Apidae, Meliponini): interspecific variation and comparison with honey of *Apis mellifera*. Apidologie 37: 480–486.

Bisui S, Layek U, Karmakar P. 2019. Comparing the pollen forage pattern of stingless bee (*Trigona iridipennis* Smith) between rural and semi-urban areas of West Bengal, India. Journal of Asia-Pacific Entomology 22(3): 714–722.

Bisui S, Layek U, Karmakar P. 2020. Utilization of Indian dammar bee (*Tetragonula iridipennis* Smith) as a pollinator of bitter gourd. Acta Agrobotanica 73(1): 7316.

Bisui S, Layek U, Karmakar P. 2021. Determination of nectar resources through body surface pollen analysis: a study with the stingless bee *Tetragonula iridipennis* smith (Apidae: Meliponini) in West Bengal, India. Sociobiology 68(3): e6173.

Brumfitt W, Hamilton-Miller JM, Franklin I. 1990. Antibiotic activity of natural products: 1. Propolis. Microbios 62: 19–22.

Campos MG, Webby RF, Markham KR, Michell KA, Cunha AP. 2003. Age-induced diminution of free radical scavenging capacity in bee pollens and the contribution of constituent flavonoids. Journal of Agricultural and Food Chemistry 51(3): 742–745.

Celemli OG. 2013. Chemical properties of propolis collected by stingless bees. In: Vit P, Pedro SRM, Roubik D (eds.), Pot-Honey: a legacy of stingless bees. Springer, New York. pp. 525–537.

Charnov EL. 1976. Optimal foraging: The marginal value theorem. Theoretical Population Biology 9: 129–136.

Chauhan A, Singh HK. 2019. Nest architecture of stingless bees *Tetragonula iridipennis* and *Lepidotrigona ventralis* in Nagaland. Indian Journal of Entomology 81(3): 613–617.

Choudhari MK, Haghniaz R, Rajwade JM, Paknikar KM. 2013. Anticancer activity of Indian stingless bee propolis: an in vitro study. Evidence-Based Complementary and Alternative Medicine 928280.

Choudhari MK, Punekar SA, Ranade RV, Paknikar KM. 2012. Antimicrobial activity of stingless bee (Trigona sp.) propolis used in the folk medicine of Western Maharashtra, India. Journal of Ethnopharmacology 141: 363–367.

Choudhary A, Singh J, Chhuneja PK. 2021. Nest Architecture and Nesting Site Preference of Smith in North-Western Plains of India. Journal of Apicultural Science 65(1): 49–59.

Chuttong B, Chanbanga Y, Sringarm K, Burgett M. 2016. Physicochemical Profiles of Stingless Bee (Apidae: Meliponini) Honey from South East Asia (Thailand). Food Chemistry 192: 149–155.

Codex Alimentarius Commission. 2001. Alinorm 41/10: revised standard for honey. Alinorm 1: 19–26.

Colucci G, De Vito V, Varricchio E, De Cunzo F, Coccia E. 2016. Identification of traceability markers in Italian unifloral honeys of different botanical origin. Journal of Nutrition and Food Sciences 6(1): 1000462.

Cortopassi-Laurino M, Imperatriz-Fonseca VL, Roubik DW, Dollin A, Heard T, Aguilar I, Venturiei GC, Eardley C, Nogueira-Neto P. 2006. Global meliponiculture: challenges and opportunities. Apidologie 37(2): 275–292.

Cunha MG, Franchin M, Galvão L, de Ruiz A, de Carvalho JE, Ikegaki M, Alencar SM, Koo H, Rosalen PL. 2013. Antimicrobial and antiproliferative activities of stingless bee *Melipona scutellaris* geopropolis. BMC Complementary and Alternative Medicine 13: 1–9.

Cunha MG, Rosalen PL, Franchin M, de Alencar SM, Ikegaki M, Ransom T, Beutler JA. 2016. Antiproliferative constituents of geopropolis from the bee *Melipona scutellaris*. Planta Medica 82(03): 190–194.

De Almeida-Muradian LB, Stramm KM, Horita A, Barth OM, da Silva FA, Estevinho LM. 2013. Comparative study of the physicochemical and palynological characteristics of honey from *Melipona subnitida* and *Apis mellifera*. International Journal of Food Science & Technology 48(8): 1698–1706.

Delfinado-Baker M, Baker EW, Phoon ACG. 1989. Mites (Acari) associated with bees (Apidae) in Asia, with description of a new species. American Bee Journal 129(9): 609–613.

Deyerling N, Achenbach J, dos Santos MM, Locher C. 2024. Physicochemical Properties, Antioxidant Activity and Phytochemical Profiling of Australian *Tetragonula carbonaria* Cerumen. Journal of Chromatography Open 6: 100164.

Divya KK, Amritha VS, Aparna B, Devanesan S. 2018. Biochemical and antioxidant properties of honey from *Tetragonula iridipennis* (Smith) of Southern Kerala. Indian Journal of Entomology 80(3): 1011–1016.

Dixon CV. 1987. Beekeeping in Southern Mexico. Conference of Latin Americanist Geographers Year Book. 13: 66–71.

Do Nascimento AS, Marchini LC, de Carvalho CAL, Araújo DFD, de Olinda RA, da Silveira TA. 2015. Physical-chemical parameters of honey of stingless bee (Hymenoptera: Apidae). American Chemical Science Journal 7(3): 139–149.

Dutra RP, Abreu BV, Cunha MS. 2014. Phenolic acids, hydrolysable tannis, and antioxidant activity of geopropolis from the stingless bee *Melipona fasciculata* Smith. Journal of Agricultural and Food Chemistry 62 (12): 2549–2557.

Flechtmann CHW, Phoon ACG, Camarcio CA. 1974. Mite associated with stingless bees from Brazil. In: Proceedings of the Fourth International Congress of Acarology, University of Austria, Sarlfelder, 315 p.

Franchin M, Da Cunha MG, Denny C. 2012. Geopropolis from *Melipona scutellaris* decreases the mechanical inflammatory hypernociception by inhibiting the production of IL-1β and TNF-α. Journal of Ethnopharmacology 143(2): 709–715.

Franchin M, Rosalen PL, Da Cunha MG. 2016. Cinnamoyloxy-mammeisin isolated from geopropolis attenuates inflammatory process by inhibiting cytokine production: involvement of MAPK, AP-1, and NF-^{k}B. Journal of Natural Products 79: 1828–1833.

Fretwell SD, Lucas HL. 1970. Territorial behavior and other factors influencing habitat distribution in birds. Acta Biotheoretica 19: 16–36.

Goulson D. 1999. Foraging strategies of insects for gathering nectar and pollen, and implications for plant ecology and evolution. Perspectives in Plant Ecology, Evolution and Systematics 2 (2): 185–209.

Greco M, Spooner-Hart R, Holford P. 2010. A stingless bee hive design for a broader climate range. Journal of Apicultural Research 49(3): 290–292.

Greenleaf SS, Williams NM, Winfree R, Kremen C. 2007. Bee foraging ranges and their relationship to body size. Oecologia 153: 589–596.

Guerrini A, Bruni R, Maietti S, Poli F, Rossi D, Paganetto G, Muzzoli M, Scalvenzi L, Sacchetti G. 2009. Ecuadorian stingless bee (Meliponinae) honey: A chemical and functional profile of an ancient health product. Food Chemistry 114(4): 1413–1420.

Habib HM, Al Meqbali FT, Kamal H, Souka UD, Ibrahim WH. 2014. Physicochemical and biochemical properties of honeys from arid regions. Food Chemistry 153: 35–43.

Hamilton KD, Czajkowski D, Kong NJ, Tran TD, Gustafson KR, Pauly G, Boyle GM, Simmons J, Steadman R, Moseley R, Brooks PR, Ogbourne SM, Russell FD. 2022. Anti-fibrotic potential of Tomentosenol A, a constituent of cerumen from the Australian native stingless bee, *Tetragonula carbonaria*. Antioxidants, 11(8): 1604.

Heard TA. 1988. Propagation of hives of *Trigona carbonaria* SMITH (Hymenoptera: Apidae). Australian Journal of Entomology 27(4): 303–304.

International Honey Commission. 2002. Harmonised methods of the international honey commission. Swiss Bee Research Centre–Liebefeld, Switzerland.

Inoue T, Sakagami SF, Salmah S, Nukmal N. 1984a. Discovery of successful absconding in the stingless bee *Trigona* (*Tetragonula*) *laeviceps*. Journal of apicultural research 23(3): 136–142.

Inoue T, Sakagami SF, Salmah S, Yamane S. 1984b. The process of colony multiplication in the Sumatran stingless bee *Trigona* (*Tetragonula*) *laeviceps*. Biotropica 1: 100–111.

Isla MI, Moreno MIN, Sampietro AR, Vattuone MA. 2001. Antioxidant activity of Argentine propolis extracts. Journal of Ethnopharmacology 76(2): 165–170.

Jayalekshmi CR. 2015. Pests and diseases of stingless bees *Trigona iridipennis* Smith (Meliponinae: Apidae). M. Sc. thesis, Department of Agricultural Entomology, Kerala Agricultural University, Kerala.

Karabagias IK, Badeka A, Kontakos S, Karabournioti S, Kontominas MG. 2014. Characterisation and classification of Greek pine honeys according to their geographical origin based on volatiles, physicochemical parameters and chemometrics. Food Chemistry 146: 548–557.

Kasote DM, Pawar MV, Gundu SS, Bhatia R, Nandre VS, Jagtap SD, Mahajan SG, Kulkarni MV. 2019. Chemical profiling, antioxidant, and antimicrobial activities of Indian stingless bees propolis samples. Journal of Apicultural Research 58: 617–625.

Kent RB. 1984. Mesoamerican stingless beekeeping. Journal of Cultural Geography 4(2): 14–28.

Kiatoko N. 2012. Distribution, behavioral biology, rearing and pollination efficiency of five stingless bee species (Apidae: Meliponinae) in Kakamega forest, Kenya. Ph. D. thesis, Pure and Applied Science of Kenyatte University, Kenya, pp. 140–226.

Klumpp J. 2007. Australian stingless bees: a guide to sugarbag beekeeping. Brisbane, Earthling Enterprises Pty Ltd, 100 p.

Kothai S, Jayanthi B. 2014. Evaluation of antioxidant and antimicrobial activity of stingless bee propolis (*Tetragonula iridipennis*) of Tamil Nadu, India. International Journal of Pharmacy and Pharmaceutical Sciences 6: 81–85.

Kumar MS, Singh AJA, Alagumuthu G. 2012. Traditional beekeeping of stingless bee (*Trigona* sp.) by Kani tribes of Western Ghats, Tamil Nadu, India. Indian Journal of Traditional Knowledge 11(2): 342–345.

Lage LG, Coelho LL, Resende HC, Tavares MG, Campos LA, Fernandes-Salomão TM. 2012. Honey physicochemical properties of three species of the brazilian *Melipona*. Anais da Academia Brasileira de Ciências 84(3): 605–608.

Layek U, Baghira NK, Das A, Kundu A, Karmakar P. 2023a. Dependency of crops on pollinators and pollination deficits: an approach to measurement considering the influence of various reproductive traits. Agriculture 13(8): 1563.

Layek U, Bisui S, Karmakar P. 2024. Flight range and resource loading-unloading behavior of stingless bee *Tetragonula iridipennis* (Smith). Journal of Apicultural Research 63(3): 400–411.

Layek U, Das A, Das U. 2022a. Floral biology, floral volatile organic compounds and floral visitors of *Chromolaena odorata*, an invasive alien species in West Bengal, India. Biodiversitas 23(4): 2118–2129.

Layek U, Das A, Karmakar P. 2022b. Supplemental stingless bee pollination in fennel (*Foeniculum vulgare* Mill.): An assessment of impacts on native pollinators and crop yield. Frontiers in Sustainable Food Systems 6: 820264.

Layek U, Das N, De SK, Karmakar P. 2023b. The botanical origin of cerumen and propolis of Indian stingless bees (*Tetragonula iridipennis* Smith): pollen spectrum does not accurately indicate latex and resin sources. Apidologie 54(2): 18.

Layek U, Das U, Karmakar P. 2022c. The pollination efficiency of a pollinator depends on its foraging strategy, flowering phenology, and the flower characteristics of a plant species. Journal of Asia-Pacific Entomology 25(2): 101882.

Layek U, Das N, Mondal R, Karmakar P. 2023c. Distribution, nesting biology, and floral preference of giant honeybee (*Apis dorsata* Fabricius) in southern West Bengal, India. In: Abrol DP (Ed.), Role of giant honeybees in natural and agricultural systems, pp. 305–323. CRC Press (Taylor & Francis), Boca Raton.

Layek U, Karmakar P. 2018. Nesting characteristics, floral resources, and foraging activity of *Trigona iridipennis* Smith in Bankura district of West Bengal, India. Insectes sociaux 65: 117–132.

Layek U, Kundu A, Bisui S, Karmakar P. 2021. Impact of managed stingless bee and western honey bee colonies on native pollinators and yield of watermelon: A comparative study. Annals of Agricultural Sciences 66(1): 38–45.

Layek U, Kundu A, Karmakar P. 2020a. Floral ecology, floral visitors and breeding system of Gandharaj lemon (*Citrus× limon* L. Osbeck). Botanica Pacifica 9(2): 113–119.

Layek U, Mondal R, Karmakar P. 2020b. Honey sample collection methods influence pollen composition in determining true nectar-foraging bee plants. Acta Botanica Brasilica 34: 478–486.

Lazarević KB, Andrić F, Trifković J, Tešić Ž, Milojković-Opsenica D. 2012. Characterisation of Serbian unifloral honeys according to their physicochemical parameters. Food Chemistry 132 (4): 2060–2064.

Marchini LC, Reis VDA, Moreti ACCC. 2006. Composicao fisico-quimica de amostras de polen coletado por abelhas africanizadas *Apis mellifera* (Hymenoptera: Apidae) em Piracicaba, Estado de Sao Paulo.Ciencia Rural. 36(3): 949–953.

Massaro FC, Brooks PR, Wallace HM, Russell FD. 2011. Cerumen of Australian stingless bees (*Tetragonula carbonaria*): gas chromatography-massspectrometry fingerprints and potential anti-inflammatory properties. Natur-wissenschaften 98: 329–337.

Molan PC. 2009. Honey: Antimicrobial actions and role in disease management. In: Ahmad I, Aqil F(eds), New Strategies Combating Bacterial Infection. Hoboken, Wiley-Blackwell, pp 229–253.

Nayak Pavithra P, Reddy Shankar M, Jayaprakash. 2013. Nesting pattern preferences of stingless bee, *Trigona iridipennis* Smith (Hymenoptera: Apidae) in Jnanabharathi Campus, Karnataka, India. International Research Journal of Biological Sciences 2: 44–50.

Nisha MM. 2002. Management of stingless bee *Trigona iridipennis* Smith (Meliponinae: Apidae) in the homesteads of Kerala. M. Sc. (Ag.) thesis, Kerala Agricultural University, Thrissur, 62–63 pp.

Nugitrangson P, Puthong S, Iempridee T, Pimtong W, Pornpakakul S, Chanchao C. 2016. In vitro and in vivo characterization of the anticancer activity of Thai stingless bee (*Tetragonula laeviceps*) cerumen. Experimental Biology and Medicine 241(2): 166–176.

Page ML, Williams NM. 2023. Honey bee introductions displace native bees and decrease pollination of a native wildflower. Ecology, 104(2): e3939.

Park YK, Koo MH, Abreu JA, Ikegaki M, Cury JA, Rosalen PL. 1998. Antimicrobial activity of propolis on oral microorganisms. Current microbiology 36: 24–28.

Pimentel RBQ, da Costa CA, Albuquerque PM, Junior SD. 2013. Antimicrobial activity and rutin identification of honey produced by the stingless bee *Melipona compressipes manaosensis* and commercial honey. BMC Complementary and Alternative Medicine 13: 151.

Pradeep S, Srinivasan MR, Saravanan PA, Balasubramani V, Harish S. 2022. Propagation of stingless bees *Tetragonula iridipennis* Smith from colonies with or without queen cell. The Pharma Innovation Journal 11(7): 480–484.

Premila KS, Devanesan S, Shailaja KK. 2013. Reduviid bug *Acanthaspis siva*- a predator of honeybee and stingless bee in Kerala. Insect Environment 19(3): 197–198.

Punith HV. 2014. Designing of suitable hives for rearing and multiplication of stingless bee, *Tetragonula iridipennis* (Smith). M. Sc. Thesis.

Pyke GH. 2015. Understanding movements of organisms: it's time to abandon the Levy foraging hypothesis. Methods in Ecology and Evolution 6: 1–16.

Raakhee M, Devanesan S. 2000. Studies on Behaviour of stingless bee, *Trigona iridipennis* Smith (Apidae: Meliponinae). Indian Bee Journal 62(4): 59–62.

Rahaman A, Das PK, Rajkumari P, Saikia J, Sharmah D. 2015. Stingless Bees (Hymenoptera: Apidae: Meliponini): Diversity and Distribution in India. International Journal of Science and Research 4(1): 77–81.

Rahaman A, Das PK, Rajkumari P, Saikia J, Sharmah D. 2018. Stingless bees (Hymenoptera: Apidae: Meliponini): diversity and distribution in India. Apidologie 39: 112–118.

Rakshitha TN. 2024. Stingless bee, *Tetragonula iridipennis* as a pollinator in bell pepper, *Capsicum annum* under protected cultivation. Internation Journal of Tropical Insect Science.

Rasmussen C. 2013. Stingless bees (Hymenoptera: Apidae: Meliponini) of the Indian subcontinent: Diversity, taxonomy and current status of knowledge. Zootaxa 3647 (3): 401–428.

Rasmussen C, Cameron SA. 2010. Global stingless bee phylogeny supports ancient divergence, vicariance, and long distance dispersal. Biological Journal of the Linnean Society 99: 206–232.

Reyes-González A, Camou-Guerrero A, Gómez-Arreola S. 2016. From extraction to meliponiculture: a case study of the management of stingless bees in the West-central region of Mexico. Chambo ED (ed.), In: Beekeeping and bee conservation: Advances in research, p. 201–223.

Reynolds AM, Smith AD, Reynolds DR, Carreck NL, Osborne JL. 2007. Honeybees perform optimal scale-free flights when attempting to locate a food source. The Journal of Experimental Biology 210: 3763–3770.

Ribeiro-Junior JA, Franchin M, Cavallini ME. 2015. Gastroprotective effect of geopropolis from *Melipona scutellaris* is dependent on production of nitric oxide and prostaglandin. Evidence-Based Complementary and Alternative Medicine. 459846.

Roopa AN, Eswarappa G, Sajjanar SM, Gowda G. 2015. Study on nesting characteristics and biology of Stingless Bee (*Trigona iridipennis* Smith.). Journal of Agriculture and Veterinary Science 8(10): 34–36.

Saaivignessh B, Manickavasagam S, Banujaa. (2023). Nesting etiquacy of stingless bee *Tetragonula* "*iridipennis*" species group. Indian Journal of Entomology 23250: e23250.

Santos HFD, Campos JF, Santos CMD. 2017. Chemical profile and antioxidant, anti-inflammatory, antimutagenic and antimicrobial activities of geopropolis from the stingless bee *Melipona orbignyi*. International Journal of Molecular Sciences. 18(5): 953.

Sen S, Borkataki S, Sutradhar P, Taye RR, Bhattacharyya B, Saranya PS, Nanda SP, Reddy MD. 2023. Pollination efficiency of stingless bee, *Tetragonula iridipennis* (Smith) on greenhouse cucumber, *Cucumis sativus* (Linnaeus). Current Science 125(8): 865–870.

Serra Bonvehi J, Escola Jorda R. 1997. Nutrient composition and microbiological quality of honeybee-collected pollen in Spain. Journal of Agricultural and Food Chemistry. 45(3): 725–732.

Shahnawaz M, Sheikh SA, Hussain M, Razaq A, Khan SS. 2013. A study on the determination of physicochemical properties of honey from different valleys of Gilgit-Baltistan. International Journal of Agricultural Science and Research 2: 49–53.

Silva AC, Gl-Santana HR. 2004. Predation of *Apiomerus pillipes* (Fab.) (Hemiptera, Reduviidae, Harpactorinae, Apiomerini) over meliponinae bees (Hymenoptera, Apidae) in the state of Amazonas, Onas, Brazil. Revista Brasileria de Zoologia 21: 769–774.

Silva IAA, Da Silva TMS, Camara CA, Queiroz N, Magnane M, De Novais JS. 2013. Phenolic profile, antioxidant activity and palynological analysis of stingless bee honey from Amazonas, Northern Brazil. Food Chemistry 141: 3553–3558.

Simoes D, Bego LR, Zucchi R, Sakagami SF. 1980. *Melalonctia sinistra* an endoparasite phorid fly attacking *Nannotrigona* (*Scaptotrigona*) *postica* (Hymenoptera: Meliponinae). Revista Brazilia de Entomologia 24: 137–142.

Sims DW, Righton D, Pitchford JW. 2007. Minimizing errors in identifying Levy flight behavior of organisms. Journal of Animal Ecology 76: 222–229.

Simuth J, Trnovský J, Jelokova J. 1986. Inhibition of bacterial DNA-dependent RNA polymerases and restriction endonuclease by UV-absorbing components from propolis. Die Pharmazie 41: 131–132.

Shanks JL, Haigh AM, Riegler M, Spooner-Hart RN. 2017. First confirmed report of a bacterial brood disease in stingless bees. Journal of invertebrate pathology 144: 7–10.

Singh AK. 2016. Traditional meliponiculture by Naga tribes in Nagaland, India. Indian Journal of Traditional Knowledge 15(4): 693–699.

Sommeijir MJ. 1999. Beekeeping with stingless bees: a new type of hive. Bee World 80(2): 70–79.

Souza B, Roubik D, Barth O, Heard T, Enríquez E, Carvalho C, Villas-Bôas J, Marchini L, Locatelli J, Persano-Oddo L, Almeida-Muradian L, Bogdanov S, Vit P. 2006. Composition of stingless bee honey setting quality standards. Interciencia 31: 867–875.

Spooner-Hart R, Halcroft M, Shanks J. 2018. Presentation on stingless bee pests and diseases. Western Sydney University, Australia.

Suntiparapop K, Prapaipong P, Chantawannakul P. 2012. Chemical and biological properties of honey from Thai stingless bee (*Tetragonula leaviceps*). Journal of Apicultural Research 51: 45–52.

Surendra NS, Bhushanam M, Ravikumar H. 2012. Antimicrobial activity of propolis of *Trigona* sp. and *Apis mellifera* of Karnataka, India. Prime Journal of Microbiology Research 2: 80–85.

Temaru E, Shimura S, Amano K, Karasawa T. 2007. Antibacterial activity of honey from stingless honeybees (Hymenoptera; Apidae; Meliponinae). Polish Journal of Microbiology 56: 281–285.

Van Veen JW, Sommeijer MJ. 2000. Colony reproduction in *Tetragonisca angustula* (Apidae, Meliponini). Insectes sociaux 47: 70–75.

Vijayakumar K, Muthuraman M, Jayaraj R. 2013. Infestation of *Carpoglyphus lactis* (Linnaeus) (Acari: Carpoglyphidae) on *Trigona iridipennis* (Apidae: Meliponinae) from India. Scholarly Journal of Agricultural Science 3(1): 25–28.

Vijoykumar K, Muthuraman M, Jayaraj R. 2013. Propagating *Trigona iridipennis* colonies (Apidae: Meliponini) by eduction method. Scholars Academic Journal of Biosciences 1(1): 1–3.

Viraktamath S, Roy J. 2022. Description of five new species of *Tetragonula* (Hymenoptera: Apidae: Meliponini) from India. Biologia 77: 1769–1793.

Vit P. 2013. *Melipona favosa* Pot-Honey from Venezuela. In: Vit P, Pedro S, Roubik D (eds), Pot-Honey. pp. 363–373. Springer, New York, NY.

Wattanachaiyingcharoen W, Jongjitvimol T. 2007. First record of the predator, *Pahabengkakia piliceps* Miller, 1941 (Reduviidae, Harpactorinae) in the stingless bee, *Trigona collina* Smith, 1857 (Apidae, Meliponinae) in Thailand. The Natural History Journal of Chulalongkorn University 7(1): 71–74.

Wesh SO, Marston RM. 1983. Nutritional Bioavailability of Zin. Washington, D. C. American Chemical Society.

Wunderink ME, Sommeijer MJ. 1996. The ultimate function of the positioning and ovipositioning process (POP) in stingless bees: an experimental study. Proceedings of Society of Experimental and Applied Entomology 7: 51–55.

Ya'akob H, Norhisham NF, Mohamed M, Sadek N, Endrini S. 2019. Evaluation of physicochemical properties of Trigona sp. stingless bee honey from various districts of Johor. Jurnal Kejuruteraan SI 2(1): 59–67.

GLOSSARY

Acetolysis: The breakdown of an organic compound using either acetic acid or acetic anhydride.

Age polyethism: The phenomena in which an animal exhibits diverse types of behaviour at different ages. It is most commonly found in social insect colonies, where different colony members undertake different jobs as they age.

Alarm pheromones: Insect alarm pheromones are chemical substances produced and emitted in response to predators to lessen the danger of predation.

Allergy: A hypersensitivity disorder of the immune system.

Anti-inflammatory: The property of a substance or treatment that reduces inflammation.

Antimicrobial: An agent that kills microorganisms or inhibits their growth.

Antioxidant: A molecule that inhibits the oxidation of other molecules, thereby reducing damage caused by oxygen or other free radicals.

Antiseptic: Antimicrobial substances applied to living tissue/skin to reduce the possibility of infection, sepsis, or putrefaction.

Apiarist: A person who keeps honeybees for their products.

Apiary: An apiary is a place where honeybee beehives are kept.

Apiculture: The keeping of honeybee colonies for commercial production of honey and other bee hive products.

Apidae: A large family of bees that contains three subfamilies: Xylocopinae, Nomadinae and Apinae.

Apitoxin: A bitter colourless liquid produced by the filamentous gland of the stinging apparatus of a worker bee.

Bee bread: A fermented mixture of plant pollen, nectar, and bee saliva that worker bees use as food for larvae and young bees to produce royal jelly.

Beekeeper: See Apiarist.

Beekeeping: See apiculture.

Bee pasturage: Honeybees gather nectar and pollen from plants as their food. The plant that yields those two substances is collectively called bee pasturage.

Bee plant: A plant that is visited by bees for nectar and/or pollen.

Bee pollen: A mass of pollen packed by worker honeybees into granules with added honey or nectar.

Beeswax: A natural wax mainly esters of fatty acids and various long-chain alcohols produced by the wax glands of worker bees.

Bioindicator: An organism or bio product used as an indicator of the quality of an ecosystem, especially in terms of pollution.

Brood: Immature stages collectively, including the eggs, larvae and pupae.

Corbicula: A smooth area on the hind tibia of certain species of bees which serve to transport pollen.

Dearth period: A period when nectar or pollen for honeybees is at a minimum in the field.

Dimorphism: The existence of two different forms of a species, especially in the same population.

Drone: Haploid fertile male bee that does not have a stinger, pollen baskets or wax glands.

Ester: An organic compound where the hydrogen in a compound's carboxyl group is replaced with an alkyl or other organic group.

Eusocial: Describing an animal society with a reproductive skew toward one or more females, cooperative brood care and overlapping generations living together.

Fatty acid: A carboxylic acid with a long side chain of hydrocarbons.

Fermentation: Energy-yielding anaerobic breakdown of a nutrient molecule, such as glucose, yields lactate, ethanol, or some other simple product.

Flight distance: A forager's journey distance from the colony to collect various resources.

Floral fidelity: Refers to the tendency of a pollinator to restrict its visits to flowers of a single species or morph.

Forage spectrum: Diversity in plant species visited by bees to collect nectar and/or pollen as principal food sources.

Foragers: Individuals who leave the nest to collect rewards from the surrounding habitats.

Foraging: Searching for wild food resources.

Foraging range: see flight distance.

Foraging trip: A forager's journey (from leaving the nest to returning to it) is known as a foraging trip or a foraging bout.

Granulation: The process of forming or crystallising into grains.

Gynes: Gynes are those destined to become queens, and these are unmated.

Honey: A natural sweet substance produced by honeybees from the nectar of blossoms, from secretions of living parts of plants, or excretions of plant-sucking insects on the living part of plants.

Honeybee: Bees of the genus *Apis* are capable of producing and storing honey and constructing perennial, colonial nests from wax.

Honeybee venom: See apitoxin.

Honeycomb: A mass of hexagonal wax cells built by honeybees in their nests to contain their larvae and stores of bee products like honey and pollen.

Honeydew: A sugar-rich sticky liquid secreted by aphids and some scale insects as they feed on plant sap.

Honey flow: A supply or period of availability of floral nectar suitable for bees to convert into honey.

Honey pots: The pots composed of cerumen are used to store honey for stingless bees.

Honey stomach: An enlargement of the posterior end of the oesophagus in the bee's abdomen where the bee carries nectar from flower to hive. As by enzymatic activities, the nectar is converted to form honey.

Honey wine: An alcoholic beverage created by the fermentation of water and honey.

Hydroxymethylfurfural: An organic compound derived from the dehydration of certain sugars. The molecule consists of a furan ring containing aldehyde and alcohol functional groups.

Hygroscopicity: The capacity of a product to react to the moisture content of the air by absorbing or releasing water vapour.

Hymenoptera: An order of insects comprising sawflies, wasps, bees, and ants, which are characterised by membranous wings.

Hypopharyngeal gland: A pair of long glands coiled in the sides of the head of young worker bees.

Immunomodulatory: An immunological change, adjustment or regulation.

Integrated pest management: Integrated Pest Management (IPM) is a strategy rooted in ecosystem principles, aimed at the long-term prevention of pest populations and their associated damage. This approach employs a variety of techniques, including biological control, habitat modification, adjustments to cultural practices, and the utilization of resistant plant varieties.

Juvenile hormone: Juvenile hormone, a primarily sesquiterpenoid hormone found in insects, is secreted by glands near the brain. This hormone is crucial in maintaining juvenile characteristics during the larval stages. It influences the moulting process, which involves the regular shedding of the exoskeleton throughout development. In adult females, the juvenile hormone is essential for the proper production of eggs.

Larva: The life stage between the egg and pupa, when bees are maggot-like.

Levy flight: Levy flights represent scale-free (fractal) search behaviours observed across various animal species.

Meliponiculture: Meliponiculture refers to the scientific practice of maintaining and managing stingless bee populations.

Meliponini: A tribe classified under the family Apidae, which includes bees that lack a functional stinging mechanism.

Melissopalynology: A branch of palynology that deals with analysing bee pollen loads and the pollen component within honey.

Metabolism: The set of life-sustaining chemical transformations within the cells of living organisms.

Nectar: A sugar-rich liquid produced in glands called nectaries, either within the flowers or by extrafloral nectaries.

Nectar flow: The period when the bees actively bring in nectar to produce honey.

Nectary: A glandular structure located outside or within a flower that secretes nectar.

Orchard: An area of land devoted to cultivating fruit or nut trees.

Parthenogenesis: A form of asexual reproduction in which the offspring develops from unfertilised eggs.

Phenolic compounds: Phenolic compounds are a category of metabolites that originate from plants' secondary metabolic pathways. This group includes polyphenols such as flavonoids, phenolic acids, tannins, lignans, and coumarins, which are naturally present in various plant products, including fruits, vegetables, cereals, roots, and leaves.

Pheromones: Pheromones are chemical substances that serve as communicative signals, influencing the behaviour of other members within insect or animal populations.

Pollen: Microgametophytes of seed plants, which produce the male gametes.

Pollen basket: See corbicula.

Pollen pellet: A pellet formed when pollen from flowers sticks to the bee's hind legs and gets mixed with nectar and saliva carried back to the hive as food supply.

Pollination: The transfer of pollen from the anthers of a flower to the stigma of the same or another flower.

Pollinator: An animal that moves pollen from the anthers to the stigmas of flowers, thus affecting pollination.

Population: All the individuals of one species in a given area whose members can interbreed.

Polyethism: Polyethism refers to the functional specialisation observed among non-reproductive members within a colony of social organisms, especially in insects.

Prebiotic: Selectively fermented ingredients that allow specific changes in the composition and/or activity of the gastrointestinal microbiota, thus conferring benefits upon host health.

Probiotics: Microorganisms that are believed to provide health benefits when consumed.

Propolis: A resinous mixture that honeybees collect from tree buds, sap flows, or other botanical sources.

Queen bee: An adult fertile female bee that lives in a eusocial bee colony.

Refractive index: The bending of a ray of light when passing from one medium into another. Refractive index n = sin i/sin r where i is the angle of incidence of a ray and r is the angle of refraction.

Round dance: A form of dance language used by successful foragers who have found food very close to the colony.

Royal jelly: A creamy product secreted by the hypopharyngeal glands of young nurse workers honey bees primarily for developing and maintaining the queen bees.

Specific gravity: The ratio of the density of a substance to the density of a reference substance.

Spermatheca: The pinhead-sized sperm-storage organ of the queen bee, connected to the vagina.

Super organism: A group of organisms functioning together as a single unit.

Surface tension: Surface tension is a contractive tendency of a liquid's surface that allows it to resist an external force.

Swarming: Swarming refers to the phenomenon in which a new colony of honey bees or stingless bees is established when the queen bee departs from the original colony and is accompanied by many worker bees.

Trophallaxis: Food or other fluids transfer among colony members through mouth-to-mouth or anus-to-mouth feeding.

Vermin: Small animals that are destructive, annoying, or harmful to health.

Viscosity: A measure of a fluid's resistance to flow. It describes the internal friction of a moving fluid.

Waggle dance: The 'figure of eight' movement performed by some honeybee foragers that conveys information about the direction and distance of a food source or other goal to other workers.

Worker bee: Sterile female bee who does not lay eggs under normal hive conditions.

SUBJECT INDEX